Ergonomic Integration in Mixed-Model Line Balancing

Lumberjack Nelson

Contents

1 Introduction **1**

1.1 Background of the Research 1

1.2 Relevance and motivation of the Research 2

1.3 Aim and objectives of the research 4

1.4 Scope of the research 5

1.5 Outline of the thesis 7

2 Literature Review **9**

2.1 Introduction 9

2.2 Worker performance using cognitive aspects in manual assembly 11

2.3 Cognitive Load Assessment for Manufacturing (CLAM) 12

2.4 Worker allocation/ Workforce planning 13

2.5 Cognitive load theory 14

2.6 Cognitive load and performance of workers 15

2.7 Cognitive penetration and Aesthetics 16

2.8 Methods to measure cognitive load 16

2.8.1 Subjective methods 16

2.8.2 Objective/Performance method 17

2.8.3 Physiological method 18

2.9 Cognitive abilities 19

2.10 Assembly line Balancing (ALBP) 20

2.10.1 Relevance of ergonomics in assembly lines 21

2.10.2 Research in ALBP 21

2.10.3 Application of MOSA as a solution to a similar optimization problem 23

2.10.4 Application of NSGA-II as a solution to a similar optimization problem 24

2.11 Summary of the literature review 25

2.12 Gaps identified in the Literature 28
2.13 Summary of Chapter 2 29

3 Problem Description and Research Methodology 31

3.1 Introduction 31
3.2 Research Problem 32
3.2.1 Influence of cognitive aspects of information and material presentation in a manual assembly environment 32
3.2.2 Identifying the cognitive abilities using virtual cognitive test 33
3.2.3 Worker allocation using personalized models from cognitive abilities 34
3.2.4 Investigation of pen assembly worker performance using wireless EEG headset 34
3.2.5 Effect of ergonomic aspects on single and mixed model assembly-line balancing problems 34
3.3 Research topics and research questions 35
3.4 Research methodology 36
3.4.1 Experimental work using a Pen Assembly Task 36
3.4.2 Experimental work using virtual cognitive task 37
3.4.3 Worker allocation using personalised models from cognitive abilities 37
3.4.4 CLAM FACTORS 38
3.4.5 Methods to measure cognitive load 40
3.4.6 Optimization Approaches – Assembly Line Balancing Problem (ALBP) 41
3.5 Summary of Chapter 3 42

4 Influence of cognitive aspects of information and material presentation in a manual assembly 43

4.1 Introduction 43
4.1.1 Cognitive perspective on manual assembly tasks 45
4.2 Environment and proposed method 47
4.3 Participants 48

4.4 Task Design 49
4.5 Experimental Methodology 52
4.6 Hypotheses 56
4.7 Results and Analysis 56
4.7.1 Statistical analysis -Kruskal Wallis test 57
4.7.2 Human Performance Measurements 59
4.8 Cognitive penetration and aesthetics 68
4.9 Development of predictive model – Logistic regression 69
4.10 Discussion on worker allocation 73
4.11 Case study: Pen recycling unit 78
4.12 Summary of Chapter 4 81

5 Identifying the cognitive abilities using virtual cognitive task 83
5.1 Introduction 83
5.1.1 Advantages of Virtual cognitive tasks 86
5.1.2 Applications of Virtual Cognitive Tasks 86
5.2 Participants 87
5.3 Experiment Procedure & Apparatus 87
5.4 Experimental design 88
5.5 Results and Discussion 89
5.5.1 Wilcoxon test 90
5.6 Summary of Chapter 5 91

6 Worker allocation using personalised models from cognitive abilities 93
6.1 Introduction 93
6.2 Methodology 95
6.3 Participants 96
6.4 Experimental design and tasks 96
6.5 Hypotheses 98
6.6 Results 100
6.7 Hypothesis for product assembly task 101
6.8 Objective and psychophysiological measurement 102
6.9 Subjective measurement 103

6.10 Workforce planning based on cognitive abilities 104
6.11 Workforce Planning methodology – Real time 112
6.12 Summary of Chapter 6 113

7 Investigation of pen assembly worker performance using wireless EEG headset 115

7.1 Introduction 115
7.2 METHODOLOGY 117
7.2.1 Participants 117
7.2.2 Experimental setup 117
7.2.3 Experimental Procedure 119
7.3 RESULTS AND DISCUSSION 123
7.3.1 Pen assembly study with 30 subjects 123
7.3.2 Pen assembly study with EMOTIV EPOC X 126
7.3.3 NASA-TLX 128
7.4 Summary of Chapter 7 129

8 Effect of ergonomic aspects on single- and mixed model assembly-line balancing problems 131

8.1 Introduction 131
8.2 Modeling methodology in an ALBP 134
8.2.1 Mathematical model 134
8.2.2 Problem instance for a single-product assembly line 140
8.2.3 Problem instance for a mixed model assembly line 142
8.2.4 Relaxation of an objective for solving the ALBP 143
8.3 Solution methodology proposed for an ALBP 144
8.3.1 Solution methodology for a single-product ALBP 144
8.3.2 Solution methodology for a mixed model ALBP 145
8.3.3 An illustrative example of the proposed solution approach using NSGA-II 145
8.3.4 An illustrative example of the proposed solution approach using MOSA 152
8.4 Results 156

8.4.1 Application of NSGA-II for a single-product ALB 156
8.4.2 Application of NSGA-II for a mixed model ALB 158
8.4.3 Application of MOSA for a single-product ALB 160
8.4.4 Application of MOSA for a mixed model ALB 162
8.5 Comparative analyses of the proposed solution approach NSGA-II and MOSA 164
8.5.1 Performance analyses 167
8.6 Discussion 171
8.6.1 Key findings 172
8.6.2 Practical implications 172
8.6.3 Study limitations 175
8.6.4 Areas for further research 176
8.6.5 Ergonomics Improvement in assembly lines 176
8.7 Summary of Chapter 8 177
9 Conclusion **179**
9.1 Introduction 179
9.2 Summary of the research 179
9.2.1 Influence of cognitive aspects of information and material presentation in a manual assembly environment 180
9.2.2 Identifying the cognitive abilities using virtual cognitive test 181
9.2.3 Worker allocation using personalised models from cognitive abilities 181
9.2.4 Investigation of pen assembly worker performance using wireless EEG headset 182
9.2.5 Effect of ergonomic aspects on single and mixed model assembly-line balancing problems 183
9.3 Contributions of the research 184
9.4 Limitations of the research 185
9.5 Scope for further research 186
187

Chapter 1

Introduction

1.1 Background of the Research

In the manufacturing industry, increasing customer expectations, stricter environmental regulations, and technological advancements have led to a growing diversity of products, with further variation anticipated in the future. This expanding product variety adds to the workload of assemblers, who must locate, retrieve and assemble numerous parts to sub parts and final assemblies. Consequently, the information provided to assemblers must be highly effective and tailored to support their tasks efficiently. However, existing information systems used in assembly processes often lack usability, as highlighted by Thorvald et al. [1]. When workers in a manual assembly confront poorly organized or presented information, their cognitive load increases because they must invest extra mental effort in searching and finding the suitable information [2]. This cognitive load can negatively impact the quality of the information processed. As an example, when time is limited, the quantity of information that a worker can effectively take in maybe constrained, leading to reduced performance. Cognitive Load Assessment for Manufacturing (CLAM) method was created to identify and mitigate cognitive load on assembly workers in a manufacturing unit [3]. From the selected CLAM factors for each workstation cognitive load assessment can be done. Galy et al. [4] reported the methods for cognitive load assessment which includes subjective, performance based and physiological methods.

As per Brolin et al. [5] an ergonomic workstation which considers cognitive aspects (information and material presentation) in a manufacturing unit can contribute to better productivity. Even after setting up the best information presentation and material presentation in a workstation, due to the individual differences of the cognitive abilities [6] of workers, the performance can be affected. To solve this a proper worker allocation can be planned according to the cognitive skills of workers. Hashemi-Petroodi et al. [7] describe workforce planning as the process of analyzing, forecasting, and planning workforce supply and demand, identifying skill gaps, and implementing targeted interventions to ensure that an organization has the best skilled workforce. Worker allocation based on the cognitive abilities of workers helps to minimise worker costs, reduce the worker count in a workstation, maximize work variability, smooth the workload, and reduce the chances of errors. To apply a scientific/mathematical approach with algorithms and to validate the experimental studies, integration of ergonomic aspects into assembly-line balancing problem (ALBP) as per Otto and Battaïa [8] helps in reducing stress over time. Incorporating ergonomic term as an objective function into an ALBP enhances productivity and minimises ergonomic concerns. The multi-objective optimization model developed by Dalle and Dini [9] can be used to consider the objective functions as skills (cognitive aspects) and energy expenditure (physical aspects). Insights into human performance can guide the workstation design of manual assembly units that are more resilient to stress, allowing assemblers to receive the right amount of information without being overwhelmed. To maintain an ergonomically friendly workplace, it is essential that workstations are user-friendly and that employers ensure workers are placed in roles suited to their skills and capabilities. Ergonomics plays a key role in this, as it considers individual differences in skills and performance, and ensures that workers with right skills are appropriately positioned at each workstations to achieve best results.

1.2 Relevance and motivation of the Research

In many industrialized countries, ergonomic aspects in manufacturing have gained attention in both research and practical applications. When workers perform repetitive assembly tasks for extended periods, they face considerable physical and mental stress. The rise in product variants has led to an increased information flow and significant information overload [10], the cognitive aspects are valuable and essential. Workstation designers in a manual assembly unit must be aware about how presenting information effectively so that assemblers can complete their tasks accurately based on the provided data. In addition to examining cognitive aspects effecting performance, it is also important to consider cognitive abilities [6] of workers and thereby allocating them to appropriate workstations.

In manual assembly, if the information provided and the materials to be assembled are not presented in such a way that allows workers to identify the objectives of their tasks, then there are higher chances of increased cognitive load. For assessing cognitive load, there are eleven CLAM factors have been identified by [3], from which the relevant factors can be selected based on the workstation/manufacturing unit being assessed. Apart from this, cognitive load can be measured using subjective methods [11], performance-based methods [12], and physiological methods [13]. A recent survey by Otto & Battaïa [8] examined research aimed at incorporating ergonomic factors into ALBP as a means of reducing cognitive load over time through optimized skill allocation. Understanding cognitive abilities of workers will allow for better balancing of assembly lines, where tasks are assigned not only based on physical capabilities but also skills.

The primary research aim is to study the effect of cognitive aspects in manual assembly unit, and based on the results, identify the performance of workers. CLAM factors considered are used to define eight combinations of information and material presentation. The cognitive load is measured using subjective (NASA -TLX), performance based (output without errors) and physiological (wireless EEG headset). The cognitive abilities of the subjects are also considered for an effective worker allocation. The role of cognitive abilities in shaping workforce allocation that align task demands optimizing human performance is also considered. An effective workforce allocation understanding of the cognitive aspects in the assembly environment and cognitive abilities of workers is reported. By integrating cognitive abilities into workforce planning and assembly line optimization, industries can enhance productivity, reduce error rates, and create a healthier work environment. The findings will contribute to have ergonomic and efficient workstations, particularly in manual assembly settings.

When workers are presented with poorly structured or overly complex information, they must exert additional mental effort to understand, process, and apply the information. This additional cognitive load can lead to confusion, slower task execution, and greater chances of errors. The performance of workers in manual assembly tasks is influenced by various factors, one of the most significant being cognitive aspects.

As industries continue to strive for greater efficiency and quality in production processes, understanding the role of cognitive abilities in human performance becomes crucial. Furthermore, an overwhelming amount of information or unclear instructions can result in mental fatigue, reducing the worker's ability to focus, remember details, and make quick decisions, ultimately impacting overall performance. Properly designed information and material presentations are crucial for reducing cognitive load, allowing workers to focus on improving efficiency, accuracy, and productivity.

1.3 Aim and objectives of the research

A lot of technical and scientific approaches are made by some advanced manufacturing units but when it comes to certain workstations with manual assemblies with variant flora of products and small parts a proper solution is difficult. Automation can be considered as a solution but due to very small parts and costs this is not feasible. Expansion of the workstation/production facility increases the investment cost but this is not practical when it comes to new products and a wide variety of products. This present research evaluates the performance of workers in the workstation based on the cognitive aspects. This research explores the effect of cognitive aspects on human performance in a manual assembly work environment, specifically focusing on presentation of information and materials, workforce planning, cognitive ability, and their impact on assembly line balance.

The objectives of this research are as follows:

- Identify and explore the applicable factors for assessing human cognitive load in manual assembly of small parts.
- Design and conduct experimental studies that vary the cognitive load and variants of the product.
- Investigate the effect of the presentation of information and material to each worker in a manual assembly.
- Explore the cognitive ability of different individuals using virtual cognitive task.
- Develop personalized models based on cognitive abilities for the allocation of work to worker.

- Validate the results of the experimental study with those of the same study carried out using a wireless EEG headset.
- Develop multi-objective optimization solution approaches, namely the non-dominated Sorting Genetic algorithm II (NSGA-II) and the Multi-Objective Simulated Annealing (MOSA), using Pareto optimality for solving a multi-objective Assembly Line Balancing Problem (ALBP) considering ergonomic aspects.

1.4 Scope of the research

When the variety of products to be assembled in a manual assembly unit increase, then cognitive load of the workers also increases. Thus, the instruction manual provided to assemblers must be highly effective and tailored to support their tasks efficiently. In a manual assembly for the assembly tasks to be performed effectively and accurately, the cognitive aspects influencing human performance have to be properly considered. In this research, the cognitive aspects such as information presentation and material presentation are taken from CLAM factors. The pen assembly task and virtual cognitive task are the experimental studies conducted in this research. A manual assembly work environment considered is a pen assembly unit where subjects will be assembling six varieties of pens. The pen assembly task was conducted with thirty subjects across eight scenarios that combine different methods of presenting information and materials. In each scenario, the participants assembled six different types of pens. In the virtual cognitive task, specific tasks/questions are associated with each cognitive ability, and scores with certain weightages are assigned based on the inputs from brain blog websites (sharpbrains). This total score of all the cognitive abilities for each subject is noted. The cognitive load measurement in pen assembly task is done using subjective, objective and physiological methods. The subjective methods used are NASA Task Load Index (NASA -TLX) [11] and Instantaneous self-assessment (ISA) which helps to investigate the cognitive load. NASA -TLX is used in both pen assembly task and virtual cognitive task whereas ISA is used only for pen assembly task. In objective method the number of pens assembled without errors and the total cognitive score from virtual cognitive task helps in identifying the performance of subjects and the cognitive load. The physiological methods which are used in pen assembly experiment are heart rate variability and electroencephalography

(EEG) [13]. The results from pen assembly task and virtual cognitive task of different subjects helps to allocate the workers based on their skills and performance in the actual manual assembly. The experimental studies (pen assembly task and virtual cognitive task) mimicked the industrial environment and a methodology is developed to allocate the workers based on their cognitive abilities. The results of the experimental study involving subjects can be applied to workers in a manufacturing unit. To provide a scientific algorithm to the experimental study assembly-line balancing problem (ALBP) is used. Assembly Line Balancing Problem (ALBP) entails allocating assembly tasks across workstations in an optimal way to ensure a balanced and efficient production process under given constraints. The worker sequence and task sequence obtained from the Pareto optimality solution of ALBP helped identify worker skills and allocate them to the workstations where they can perform best.

The information and material presentation in a manual assembly must be an aid in reducing the cognitive load of the manual assembly worker. A good set of instructions and the workstation setup with the components arranged ergonomically considers both physical and cognitive aspects of workstation design. The cognitive abilities contribute to worker performance and it is necessary to identify them in any of the tasks in a production unit and thereby improving the productivity and rework caused due to errors. Incorporating ergonomic considerations into an ALBP enhances productivity [9]. This research can be used as a methodology for allocating the workers based on their skills in similar manual assembly manufacturing units.

The schematic representation of the research plan of the present research is outlined in Figure 1.1.

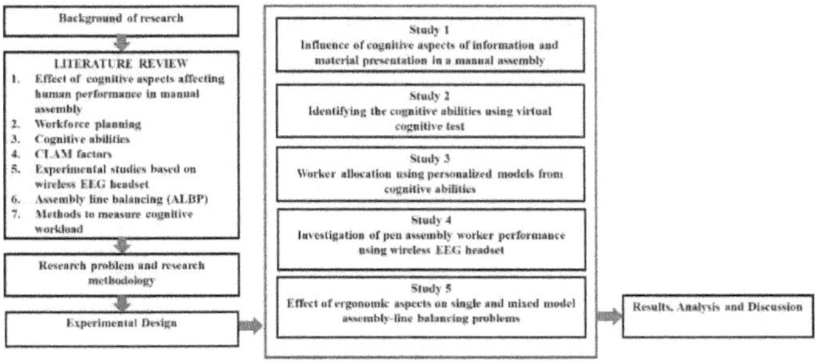

Fig. 1.1 Research plan of the present research

1.5 Outline of the thesis

This thesis documents evaluation of manual assembly environment using cognitive aspects: experimental and model-based approach. The thesis is organized into nine chapters, as described below. This present chapter introduces the background of the research by portraying the cognitive aspects affecting human performance in manual assembly. The chapter includes the motivation for the research, followed by the objectives and scope of the study.

Chapter 2 presents a systematic review of the literature relevant to the research domain. The gaps identified in the current literature are also discussed.

Chapter 3 describes the research problem and the research methodology adopted in the present research.

Chapter 4 describes the first study, as Influence of cognitive aspects of information and material presentation in a manual assembly

Chapter 5 identifying the cognitive abilities using virtual cognitive test

Chapter 6 explains worker allocation using personalized models from cognitive abilities

Chapter 7 discusses investigation of pen assembly worker performance using wireless EEG headset

Chapter 8 explains effect of ergonomic aspects on single- and mixed model assembly-line balancing problems

Chapter 9 summarizes the major findings, contributions to the literature, limitations in the present research, and the scope for further research.

The list of references and the publications from the present research are provided at the end.

Chapter 2

Literature Review

2.1 Introduction

Manufacturing different products in this modern era is more concentrated on transforming from mass production to mass customization, necessitating systems that can adapt to changing customer demands, emerging technologies, and environmental considerations [14]. According to Hu et al. [14], this transition is challenging, especially when aiming for low-cost solutions. To stay competitive, manufacturers are creating production systems that are flexible and efficient. As a real-time example, an automotive industry in Sweden utilizes mixed-model assembly, enabling the production of a variety of vehicle models on one assembly line. This approach results in a diverse array of products that must be assembled. Despite increasing automation in vehicle manufacturing, manual assembly remains essential and must be taken into account. Mixed-model assembly lines handle both high-volume products and customized variants simultaneously, which complicates the assembly process [15]. There is a significant risk that workers will use utilize prior assembly patterns, employing automated responses which leads to errors. From an ergonomics perspective, this arrangement can be demanding on workers, who may not be mentally prepared for the variety of products to be manufactured. They may also need to look for the instructions to assemble and the components, which adds complexity to their tasks. To solve this issue a well-defined information presentation is necessary for the workers in a manual assembly to achieve effective performance [15]. The topics considered in the literature review are summarized in Figure 2.1.

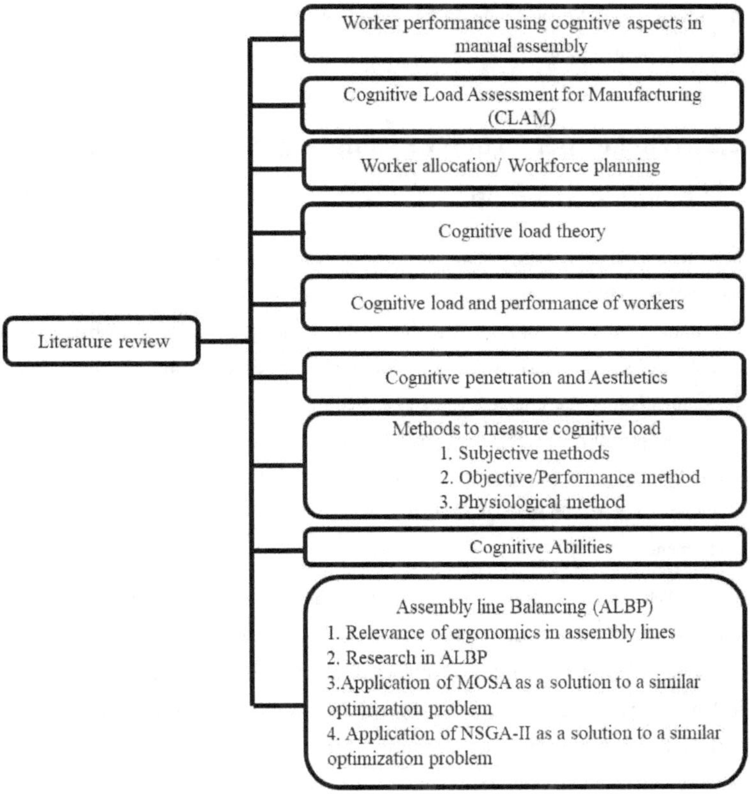

Figure 2.1 Taxonomy of literatures

The chapter is organized as follows: Section 2.2 reports the literature review of worker performance using cognitive aspects in manual assembly. Section 2.3 describes the literature review of Cognitive Load Assessment for Manufacturing (CLAM) and worker allocation/ workforce planning in section 2.4. Section 2.5 includes literature review of cognitive load theory and section 2.6 explains cognitive load and performance of workers. Section 2.7 and 2.8 explains the cognitive penetration and aesthetics and methods to measure cognitive load respectively from the literature review. Section 2.9 explains the literature review of cognitive abilities whereas section 2.10 explains assembly line balancing problem. Section 2.11 described the summary of the literature review. Section 2.12 enumerate the gaps identified in the literature review. The chapter is concluded in section 2.13.

2.2 Worker performance using cognitive aspects in manual assembly

In a manual assembly task, the worker will be having a higher cognitive load due to the increase in product variants. The various cognitive demands in manual assembly, include the amount of knowledge required, handling a variety of parts available, system complexity, and physical layout of workstations. When time pressure due to time constraints is low, these factors can be handled relatively easily, but when it comes to high time pressure, the cognitive load is increased. [16-18] identified that product variability has a negative impact on consistency and competitiveness, as well as ergonomic aspects of manual assembly work environment and reduces the overall performance.

When product varieties increase, workers are frequently faced with an increasing variety of parts and sub-parts to be assembled at workstations. The kitting theory is an interesting area when this topic is considered, which refers to the supply of kits which includes the parts and subparts to be assembled for worker convenience. The worker at each workstation must be able to identify the correct parts to be assembled on each product variant in mixed-model assembly. Continuous supply provides an uninterrupted flow of supplies instead of grouping into kits. Kitting can promote learning, and as a result, learning times are reduced and product quality is improved. With an ergonomics approach, the advantage is that the worker just needs to be concerned about how to assemble, rather than what parts to assemble, which improves quality of product [19].

Human performance in manufacturing was considered in previous studies by Hanson and Brolin [20] and Brolin et al. [5] in which it is clear how information and materials are presented to the worker. A kitting concept is used for material presentation. These structured and unstructured kits were modified according to the requirements for a manual assembly of a pen by considering cognitive aspects, time, and worker error. There are four different types of material presentation used in this study. It was therefore of interest in the experimental analysis to see whether various types of instructions had an impact on the assembler. The recommendation was to use text and number instructions, to present information in manual assembly, as well as photograph instructions, to test a more visual alternative, based on previous works by [20-22]. Eight combinations considering both information presentation and material presentation are considered in this study.

As per the literature, to improve ergonomics at workplace, human-robot collaboration can be introduced. Although technology supporting worker-robot collaboration is fast evolving, modifying workstation designs based on these improvements is lagging behind. Stork et al. [23] discussed various cognitive theories that are relevant to manual assembly, including attention, perception, memory, and problem-solving. The present research aims to provide a guidance on how to incorporate cognitive aspects (information and material presentation) while designing a workstation and allocate workers according to their performance.

2.3 Cognitive Load Assessment for Manufacturing (CLAM)

Mattsson et al. [24] and Mattsson et al. [25] developed a usability approach to investigate a workstation and to support workers in handling complexity in manufacturing. They used elements or factors gathered from a variety of other methods from the literature review. The elements or factors taken into account includes product varieties, work environment, layout, tools and work instructions.

Lindblom and Thorvald [3] developed a useful framework and a modified form of usability approach for considering cognitive aspects related to manual assembly have recently been proposed. CLAM method was proposed by Thorvald et al. [26]. with the aim of identifying and reducing possible cognitive load of manual assembly workers. It was noted that by proactively identifying relevant concerns at manual assembly workstations, time and resource savings can be achieved on the shop floor. Workstation designers should create workstations that help minimise cognitive load. Based on manual assembly environments, there are eleven CLAM factors to be considered. These CLAM factors are valuable inputs to the reduce cognitive load of worker in a manual assembly.

Undertaking operations in workstations which has different product variants frequently involve complex jobs and poor performance from workers, which lead to human mistakes and serious consequences. The research conducted by Ghalenoei et al. [27] focused on determining the cognitive performance of control room operators, while this study examines the cognitive load experienced by manual assembly workers. The study identifies the CLAM factors for the pen assembly task out of the 11 factors and the individual performances of workers.

2.4 Worker allocation/ Workforce planning

According to various studies conducted earlier, the task processing time for a single worker is considered fixed. The processing time of a job, according to Battaïa et al. [28] also depends on the workforce. The importance of workers in contributing to overall productivity must be included during the initial stages of workstation design. Workforce planning is often combined with overall design, planning and scheduling of the manufacturing unit. In most cases, the physical layout and the elements of the manufacturing and transporting equipment is determined before the workforce is planned, which is not recommended [28]. The complexity of tasks to be performed on a particular workstation and the skill required to complete that task must be planned in the initial stages of the workstation design. Workforce planning helps to optimize performance requirements in various manufacturing systems: minimise cycle time, costs of equipment's, and the number of finished products per time unit. The workers following regular shifts are assigned at random orders without considering the work constraints and cognitive aspects. The present study considers these factors to allocate the workers to each workstation. Moreover, Hashemi-Petroodi et al. [29] documented studies concerning workforce reconfiguration strategies within the context of workforce planning across different production environments. It is reported that the increase in mass customization and reduced product lifecycles requires skilled labour to avoid the human errors and rework. Furthermore, the literature review shows that there are only a limited number of real-time studies on worker allocation in manual assembly.

The influence of cross-trained workers, as well as learning effects, on the performance of manufacturing systems was examined by Chu et al. [30]. Wang et al. [31] investigated a semi-automated automobile engine assembly line where the previous approach of using fixed personnel in each manual assembly section is replaced by a new strategy that employs walking workers. The primary issue with this design is the necessity for cross-training each worker so that they can attain a satisfactory skill level for the assembly tasks assigned to them. This work identifies the workers who need training and specifies the kind of training such workers to fit them in a particular workstation. In this study, the performance of workers is studied making them perform certain tasks in a similar work environment to that of a real-time manual assembly. This work contributes to the idea of training and individual differences which affect the productivity in manufacturing units. The employers can conduct a pilot study like the pen assembly study to identify the individual performances and allocate workers to appropriate workstations.

2.5 Cognitive load theory

Cognitive Load (CL) theory, which categorizes CL into intrinsic, extraneous, and germane types, was explained by Sweller [32]. Sweller's cognitive load theory suggested that high mental workload would require the individual to allocate extra resources to entering information, and that the demand for extra resources may reduce processing efficiency and performance. Intrinsic CL is the load caused by the task complexity of the information, extraneous CL is referred to as the load imposed by way of information presented to the users, and germane CL refers to the effort in the development of new schema (internal representation based on learning) in users in the process of learning new ways to complete tasks. This research considers all the three loads of cognitive load theory and applies it as subjective measurement considering the proposed eight scenarios. The eight scenarios with combinations of information and material presentation leads to different extraneous CLs given the same level of task complexity (intrinsic CL) and similar levels of prior learning (germane CL) for workers.

Cognitive load can be a result of an interaction between task demands and worker characteristics. Schnotz and Kürschner [33] proposed that germane load would correspond to the "conscious application of learning strategies (i.e., strategies, which are not automated yet), conscious search for patterns in the learning material in order to deliberately abstract cognitive schemata (i.e., mindful abstraction) in order to complete a task more easily. As a result, this paper identified the importance of the three cognitive loads using subjective measures for each worker in order to justify their allocation based on the number of pens assembled.

2.6 Cognitive load and performance of workers

As per Sweller, cognitive load evolved from Cognitive Load Theory [34-35]. Cognitive load theory defines cognitive load as the demands being put on the storage and processing of information in the human working memory [33]. Chen et al. [36] explained that cognitive load is the number of working memory resources used. In ergonomics and human factors, cognitive load is referred to as mental workload, mental effort, or mental demand and is defined as the amount of mental activity required to perform a task e.g., Young et al. [37] and Van et al. [38]. The definitions of cognitive load differ according to the environment or the terms being considered, but the concepts of working memory and capacity are considered.

Cognitive load has been studied in various scientific domains like cognitive psychology, instructional design, human factors, and ergonomics. Cognitive load plays an important role in how well human cognitive processes function, as well as in psychological well-being. Numerous studies identified the association between cognitive load and the functioning of cognitive processes Chen et al. [36] suggested that cognitive overload (high levels of cognitive load) lead to a negative impact on the performance of working memory. Similarly, cognitive underload (low level of cognitive load) results in worst performance due to a lack of motivation or boredom. A balance between cognitive load and the cognitive process is required for better human performance to improve productivity and quality of industrial processes (in the manual assembly context).

Cognitive load is relevant to cognitive performance and well-being. In this study, a production unit with manual assembly task is used as an example. In industry, the cognitive load of workers during assembly tasks is likely to depend on the complexity of the task at hand, the work instructions provided, and the experience of the worker. This study considers all three factors by varying the varieties of products. The increase in the number of product varieties, their number of parts, and the introduction of new products led to increasingly complex processes [39].

In this complex industrial environment, workers spend time on repetitive processes as these cannot be fulfilled by machines. From a cognitive perspective, the workers will be learning to use new products and, thereby, parts when the varieties change according to the demand. Some workstations in the industries may result in more information processing (Identifying the assembly of product varieties with small parts makes it a complex process) and information storage (It is necessary to remember the products, parts, and their classification and how they are assembled). Noroozi et al. [40] stated that with different physiological signals measured, a correlation of these signals with cognitive load, attention, engagement, etc., cannot be interpreted or analysed.

2.7 Cognitive penetration and Aesthetics

Burnston [41] claimed that the cognitive penetration and aesthetic engagement has a secure relationship. Cognitive penetration refers to the idea that our perceptual experiences might be influenced by a person's beliefs, expectations, and other cognitive states, while aesthetics explores the nature of beauty, art, and taste, and how they relate to human experience. A real time example is that, imagine believing someone is angry, which makes you perceive their facial expression as expressing anger, even if it's not objectively true. In this study, the various pens used have some aesthetic influence, helped the subjects to select certain pens at first, but the number of parts of that aesthetically attracted pen had more number of parts to be assembled which leads to cognitive penetration and resulted in errors.

2.8 Methods to measure cognitive load

Several researchers reported their attempts to measure the cognitive loads, and certain methods were also mentioned. The methods for cognitive load calculation include physiological measurements [42-43], task performance measures [44], and subjective assessment measures [45]. This paper considers subjective methods, performance-based methods/objective methods and physiological methods.

2.8.1 Subjective methods

Asking individuals to report on their perceived workload, e.g., through self-report questionnaires or rating scales. Although subjective ratings are easy to obtain and straightforward, they can be influenced by personal biases. This study used NASA Task Load Index (NASA-TLX) and Instantaneous self-assessment (ISA) methods.

In ergonomics research a commonly used method is NASA Task Load Index (NASA-TLX) which is proposed by Hart and Staveland [11] which is a tool that measures how much mental and physical work is required to complete a task. It's a subjective rating scale that can be used to assess the effectiveness of a task, system, or team. NASA-TLX is a workload evaluation tool used to investigate the expected workload. Subjects rate their perceived workload on six subscales which includes exploring the Mental Demand, Physical Demand, Temporal Demand, Own Performance, Effort, and Frustration Level. A weighted average of the ratings on the subscales is used to calculate an overall workload score using excel macros file.

The second subjective method used in this study is ISA, which is a technique that has been developed as a measure of workload to provide immediate subjective ratings while performing the task. ISA is a self-rating method of the current mental workload during a task on a scale of 1 (low) to 5 (high). It is normally queried every 2-5 minutes during the activity. Kirwan [46] used the following ISA scale to assess air traffic controllers (ATC) workload.

2.8.2 Objective/Performance method

In performance-based method, certain outcomes are evaluated such as error rates, reaction times, and task completion times, to infer cognitive load. Comparing the same performance measures under different levels of task complexity assess cognitive load.

Evaluating outcomes, such as error rates, reaction times, and task completion times, helps to measure cognitive load. In this research the subjects cognitive load is inferred from their performance based on the total number of pens assembled without errors. Chi and Lin [12] demonstrated a trade-off between these performances criteria, as the time needed to complete a task increased when accuracy requirements increased, whereas a decrease in accuracy occurred when task rapidity requirements increased. Mental load in turn would indicate the portion of cognitive load that is imposed exclusively by the task and by environmental demands [47-48]. Several results from different investigations demonstrated that as the cognitive load increased, worker performance decreased, while it subsequently improved with training.

2.8.3 Physiological method

In physiological methods, physiological responses, such as heart rate variability (HRV). electroencephalography (EEG) [13], are measured to indicate cognitive load. Changes in various bodily processes and states have also been reported with changes in mental workload. One major advantage of psychophysiological measures is the continuous availability of bodily data, allowing load to be measured at a high rate and with a high degree of sensitivity, even in situations in which overt behaviour is relatively rare and explains about heart rate as a psychophysiological measurement [13]. Psychophysiology is the combination of psychology and physiology. Psychology is the study of the human mind, the brain's functions, and human behaviour. Physiology is the study of the normal functions of living organisms and their biological parts. However, psychophysiological measures are also very sensitive to physical effort and will reflect specific mental load variations only for activities involving little or no physical effort [49].

Cranford et al. [50] suggest that heart rate can potentially serve as a measure of changes in cognitive load and that it is suitable to examine the data by framing hypotheses. According to Jercic et al. [51] an underlying hypothesis states that heart rate is directly linked to physical activity but is also a measure of both the sympathetic and parasympathetic autonomic nervous system activity, revealing the indirect link of heart rate with cognitive load. Heart rate is also related to physical activity and emotions. Heart rate variability is often used as an indication of the modulation of the autonomic nervous system. Solhjoo et al. [52] tried to explain that stress or arousal will cause an increase in blood pressure and a decrease in heart rate variability, thereby identifying a link with cognitive load. Vanneste et al. [53] reported that cognitive load is related to physiological data (EEG), which enables it to measure cognitive load in real-time and without interrupting the subject through directly measurable variables.

The second physiological method used is electroencephalography (EEG) in which a wireless EEG headset (Emotiv Epoc X) is used to measure the cognitive load. The effects of mental fatigue on cognitive capacity are concentrated in some studies. Xiao et al. [54] proposed a multi-modal approach to detect and estimate mental fatigue. They utilized physiological signals such as electroencephalography (EEG) and heart rate variability (HRV) to capture the mental states of assembly workers. Xiao et al. [54] reported that the proposed multi-modal approach could effectively detect and estimate mental fatigue in the manual assembly process. An experiment in which participants performed a manual assembly task while their brain activity was recorded using electroencephalography (EEG). The participants were given different types of instructions, including explicit instructions that provided clear guidance on how to perform the task and implicit instructions that required the participants to figure out the task on their own.

Cognitive load is also assessed during multitasking using single-channel EEG and graph methods. The aim was to address the challenge of measuring cognitive load in complex multitasking scenarios, which can have implications for performance and user experience. A novel approach that involved analysing EEG data from a single electrode and applying graph-based methods to assess cognitive load. Boring et al. [55] developed a method for continuously decoding cognitive load using EEG data and identifying both task-general and task-specific correlates of cognitive load. The study aimed to explore the relationship between cognitive load and EEG signals across different tasks. Antonenko et al. [42] investigated that how EEG can provide insights into the cognitive processes underlying learning and problem-solving tasks. The previous research studies were reviewed which used EEG to measure cognitive load.

Fouad and Labib [56] have used Emotiv EPOC X neuroheadset to acquire data which helped this study to identify the methodology of data acquisition and processing. To make the data acquisition easy and to read the data (EmotivPro Software) based on the experimental setup of the current study, commercially available equipment Emotiv EPOC X wireless EEG headset is used.

2.9 Cognitive abilities

The 'cognitive revolution' was initiated more than half a century ago, through a multidisciplinary concept called cognitive science, the study of mind. A recent study suggests that cognitive science is heavily dominated by psychology Nunez et al. [57]. In the cognitive neuroscience of executive function, studies of individual differences have become increasingly important. The emphasis is primarily on the impact of individual skills and individual performances which helps in worker allocation. The focus is on skill set as a personality dimension within normal populations rather than within clinically anxious ones, and there is a focus on individual variations in cognitive skills as per Eysenck [58].

Cognitive scientists work to comprehend the nature of our thinking, what aspects are universal and which are variable, and how the diverse situations we face react to and shape human cognitive mechanisms [59]. Cognition is a term that refers to the mental processes involved in learning and understanding. Many aspects of intellectual functions and processes, such as attention, memory, computation, decision making and learning are also covered in cognition. Cognitive processes employ and generate new information based on pre-existing knowledge.

Increasingly, cognitive neuroscientists agree that ongoing progress in understanding human cognitive skills would require not only new data acquisition, but also data synthesis and integration through studies and laboratories. Efforts for an increased focus on cognitive skills and their effective way of identification of tasks and experimentation contributed to the experimental design of the study as per Yarkoni [60]. Fernandez [61] explained about the different cognitive abilities and how these cognitive abilities are used for measuring human performance. Cognitive abilities are brain-based abilities required from the easiest to the most complex to carry out any task. According to Zhang [6], cognitive abilities are linked to the processes of reading, recalling, problem-solving etc.

The cognitive abilities & skills involved are given certain scores based on the cognitive tests which are done through brain blog research websites which integrate brain health into training, leadership programs etc.

2.10 Assembly line Balancing (ALBP)

The Assembly Line Balancing Problem (ALBP) addresses the optimal distribution of assembly tasks among sequential workstations, aiming to establish a balanced and efficient production line within predefined constraints.

2.10.1 Relevance of ergonomics in assembly lines

According to Naeini and Mosaddad [62], ergonomics aims to provide a workstation that balances worker safety, wellbeing and productivity. The primary goal of any manufacturing unit is to increase profits; traditional assembly line optimizations only consider economic factors, such as the production rate. Otto and Scholl [63] introduced ergonomic risks into ALB, in which the indirect costs incurred owing to worker health issues, such as medical costs and absenteeism, are considered. If a skilled worker is absent, a replacement is required to run the production line; however, productivity is reduced because of the new worker's learning process and slow performance.

Village et al. [64] explored the ergonomics of assembly line manufacturing and the limitations of implementing ergonomic principles to improve assembly quality. A scorecard and rating system was created as part of this research. Two ergonomic approaches were developed for quality improvement during the initial stages of design and assembly line ramp-up. Caputo et al. [65] designed a framework based on a concurrent engineering method that utilized the digital twins of the Fiat Chrysler automobile assembly line. This framework helps reveal ergonomic problems on an assembly line, and design issues are solved in early stages of design. With this method, significant cost savings for the repair of design flaws can be realized, as well as a significant reduction in the time to market. Based on these studies, it becomes evident that ergonomic considerations are essential when treating ALBP. Nonetheless, it is the responsibility of every employer to simultaneously protect employee health and wellbeing and improve production rates.

2.10.2 Research in ALBP

ALBP has attracted significant research attention with a mathematical formulation of an ALB model. For a Simple Assembly-Line Balance Problem (SALBP), heuristic solution procedures were developed and recorded in a review by Scholl and Becker [66].
There are three main trends of ALBP identified from literature (i) Ergonomic risk(s) of a workstation are assessed, based on the tasks assigned, by adopting an existing method that is limited by a predefined maximum ergonomic risk value. This methodology has been previously reported by Otto and Scholl [63]. (ii) To enhance the ergonomic

situation, an ergonomics term can be included in the objective function [67]. (iii) The ergonomic aspects are combined with mixed-model production, worker assignment [68] etc. Of the three aforementioned trends, the second and third bear more resemblance to the present research. Mixed-model ALB with a worker assignment based on skills is also performed as per the third category listed above.

Samouei and Ashayeri [69] solved a multiproduct ALBP in a straight assembly line layout, whereas in this study, both single- and multiproduct ALBP were considered for a straight assembly line layout. For small-scale production units, both single- and multiproduct ALBP must be solved for better results, as there will always be small subassembly lines where only a single product is being manufactured, which has to be linked to the main assembly line. Cil et al. [70] solved a multiproduct ALBP problem considering a single objective, whereas this work has three-objective functions in a straight assembly line layout. Dalle and Dini [9] used a weighted-sum approach to address multi-objective ALBP. The drawback of this approach is already mentioned above and this study mitigates it by utilizing the Pareto optimality method.

Assembly line construction is a strategic and high-cost choice that is critical for properly designing an assembly line and balancing the workload at workstations. The advantages of line balancing include enhanced efficiency and productivity in the manufacturing units. Eghtesadifard et al. [71] explained that when the number of workstations is optimized for a given cycle time in the classical model of ALB, it is known as an SALBP of type 1 (SALBP-1), whereas minimizing the cycle time for a given number of workstations is known as an SALBP of type 2 (SALBP-2). The gap between the SALBP-1 model and the planning of real-time assembly line configurations with all the constraints is vast. Mass producers utilize assembly lines in various industries, including the production of various products such as electronic goods, automobiles, and airplanes [72]. Automobile assemblies are undoubtedly the most widely used components. Hundreds of employees must assemble thousands of parts and coordinate with machines at several workstations on a typical automotive assembly line, with a cycle time ranging from 60 to 90 s [73].

To account for the constraints related to the problem parameters, Ozdemir et al. [67] introduced a fuzzy multi-objective problem for assembly line balancing, which was evaluated using robust optimization approaches and ergonomic risks. Pilati et al. [74]

introduced the ALBP as the best method to assign numerous operations, leading to various assembly line processes. Weckenburg and Spengler [75] considered an ALBP for cost effective system configurations, were robots and human workers will be allocated to a production line. Boysen et al. [76] reviewed previous studies on ALB published in 2007. This review reports a methodology for retrieving precedence data, reviewing new problem variants, and developing models of ALB extensions as adopted in the literature. Anel et al. [77] investigated a mixed-model ALBP type 2 with single-sided linear assembly line configurations. These are the typical configurations adopted in small- and medium-sized industrial environments.

According to the literature review, ALBPs are categorized based on their objective functions and problem structures. Based on the objective functions, ALBPs are named type F, type E, and types 1, 2, 3, 4, and 5. Similarly, based on the problem structure, ALB models are classified as single-model ALBPs or multi-model (mixed model) ALBPs. However, this study addressed an ALBP type F from the objective function category. Our study implemented an ALBP solution that minimised the number of workstations, the total skill level required of assigned workers, and the energy expenditure variance of workers among workstations.

2.10.3 Application of MOSA as a solution to a similar optimization problem

This subsection summarizes the application of the MOSA algorithm. Varadharajan and Rajendran [78] explored the problem of permutation flow shop scheduling to minimise the makespan and total flow time of jobs. MOSA was proposed as a solution, wherein two initial sequences were obtained using local search heuristics. A comparative study was presented between MOSA and multi-objective genetic local search (MOGLS), an elitist, non-dominated sorting GA, and gradual priority weighting. Finally, it was reported that the proposed MOSA required less computational effort.

Khan and Govindan [79] conducted a study to minimise the objective functions, makespan, and maximum tardiness of jobs, which was considered a multi-objective permutation flow shop problem. They developed and evaluated the MOSA with random problems. The efficiency of MOSA was compared with that of different multi-objective genetic algorithms, such as the strength Pareto evolutionary algorithm II (SPEA-II) and

MOGLS. According to the computational results, MOSA outperformed SPEA-II and MOGLS.

Considering the computational efficiency of MOSA, most researchers have adopted this approach to solve problems in different domains. Cakir et al. [80] applied the MOSA technique to solve ALBPs at parallel workstations. Lara et al. [81] worked on a multi-objective meta-heuristic algorithm based on simulated annealing for political districts, and a comparison study was performed using 12 real datasets. Similarly, Dinh et al. [82] explained the ALBP in the garment industry using simulated annealing.

Fang et al. [83] used a mixed-integer evolutionary simulated annealing algorithm was developed to minimise cycle time, peak workstation energy consumption, and total energy consumption. Computational experiments suggested that the MOSA outperformed other multi-objective algorithms in terms of solution quality and computational efficiency.

Therefore, as per the review of related literature, MOSA has been reported as a computationally efficient approach to solving combinatorial optimization problems. The MOSA can be applied to construct Pareto-optimal solutions that are non-dominated in the explored solution space.

2.10.4 Application of NSGA-II as a solution to a similar optimization problem

A brief review of earlier attempts to generate non-dominated solutions using a non-dominated sorting GA is presented in this subsection.

Murata et al. [84] applied a multi-objective genetic algorithm (MOGA) to a bi-objective flow shop scheduling problem to determine non-dominated solutions. Furthermore, the model was extended to address multiple objectives. Interestingly, Ishibuchi and Murata [85] extended MOGA by incorporating a local search procedure into the offspring, and the algorithm was termed MOGLS. Srinivas and Deb [86] proposed a non-dominated sorting GA. Chang et al. [87] proposed a gradual priority weighting (GPW) approach-based GA.

Multi-objective evolutionary algorithms (EAs) that adopt non-dominated sorting and sharing have drawbacks such as computational complexity, non-elitism, and the requirement to define a sharing parameter. Deb et al. [88] suggested a non-dominated sorting-based multi-objective EA (MOEA) called NSGA-II, which alleviates all three drawbacks.

According to Yusoff et al. [89], the computational performance of an algorithm depends on the optimization approach adopted within the algorithm. Consequently, several researchers have proposed improvements to the NSGA-II algorithm to address some of its flaws while solving optimization problems. Rajkumar et al. [90] stated that NSGA-II is a fast MOEA that uses an elitism-preserving approach to implement elitism in multi-objective searches. The quality of convergence toward a genuine Pareto-optimal set is improved by elitism.

NSGA-II is an upgraded version of Deb's non-dominating sorting GA, first described in 2001 [91]. Over the years, NSGA-II has proven to be a very good MOGA used by researchers to solve many multi-objective optimization models. Researchers have applied various methods to handle multiple objectives, such as the weighting method [92].

By treating the function as a bi-objective problem and ignoring one objective in each run, NSGA-II obtains a Pareto-optimal front for solving three-objective optimization issues. The literature shows that NSGA-II can yield a better Pareto front in multi-objective optimization problems. Moreover, a dimensionality-reduction approach can be applied to simplify the computational process by fixing one of the objectives among the multiple objectives considered.

2.11 Summary of the literature review

The main findings of literature review were that the cognitive aspects of human performance in manual assembly have recently attracted greater interest, that research on this topic remains scarce. The human performance and worker allocation was of key importance while studying the manual assembly. The review of literature suggests the keywords such as the cognitive aspects, workforce planning, cognitive abilities, CLAM factors, experimental studies based on wireless EEG headset and ALBP. A consolidated table with the summary of literature review from previous studies is depicted in Table 1.

Table 2.1 Summary of literature review

Reference	Research Focus	Research findings
[41]	Aesthetic differences vary the productivity when the varieties of products are more.	Cognitive penetration and Aesthetic considerations are not considered in recent studies.
[16]	Ten subjects assembled hacksaws based on the instructions provided. In their initial learning, subjects employed a variety of strategies and typically converged on one pattern. Participants used the strategy of selecting and putting together the heavier and longer components first, and then adding the smaller and lighter ones.	Even when it came to a simple product, subjects demonstrated various assembly patterns. The data suggests that it is necessary for the to ensure a proper information presentation and material presentation in workstations.
[5]	Experimental study of cognitive aspects influences the worker and thus effects the productivity and output.	Application of this influence of cognitive aspects are not utilized for skill-based worker allocation.

Table 2.1 (Contd.)

Reference	Research Focus	Research findings
[3]	The proposed framework aims to proactively address and describe issues related to cognitive load by appropriately designing the work station and assembly task.	Neglecting cognitive and user perspectives in manufacturing entails significant costs.
[26]	CLAM method and the factors are described.	Many studies have explored cognitive aspects, skill allocation, workload assessment in an individual manner, but a collaborative approach including cognitive aspects is missing.
[93]	The application of cognitive ergonomics in work environments, worker well-being, and overall productivity are explained.	Cognitive load will be reduced and worker performance will be improved by incorporating cognitive ergonomics
[9]	This paper proposes a multi-objective genetic algorithm aimed at addressing the ALBP based on energy expenditure.	Skilled workforce is important and needs to be considered in manual assemblies.
[58]	The focus is on skill set as a personality dimension within normal populations rather than within clinically anxious ones, and there is a focus on individual variations in cognitive skills.	Ergonomics aspects in assembly lines were based on physical ergonomics, whereas skill-based studies on worker performance are limited.
[53]	Multimodal physiological data has certain limitations and the classifiers are not properly identified.	Many advanced multimodal approaches measure cognitive load using physiological measures like EEG.

2.12 Gaps identified in the Literature

The primary findings from the literature indicate a recent surge in interest regarding the cognitive aspects of human performance in manual assembly. However, the body of research on this subject remains limited. The following research gaps have been identified through a review of the related literature.

- Many studies have explored the cognitive aspects of manual assembly, skill-based allocation, and cognitive workload individually, but research integrating all three concepts within manual assembly is scarce.
- The human temporal variability is evolving, but it has shortcomings in terms of understanding how the physical and cognitive aspects of human response interact over time. There is insufficient research on how individual cognitive abilities impact manual assembly tasks, particularly when tasks require a high cognitive load.
- Much of the research focusses on pre-task cognitive evaluation, while there is a lack of real-time measurement of cognitive load and performance. The use of wireless EEG devices could provide information on the dynamic cognitive load during task performance.
- Many studies in the literature define cognitive abilities, but there is no experimental investigation to determine cognitive abilities, particularly in a manual assembly task.
- A comparison of performance based on subjective, objective, and physiological measurement methods with cognitive abilities is also unexplored.
- Current workforce planning models focus primarily on physical aspects, such as shift patterns or task distribution. There is a gap in understanding how cognitive aspects should be integrated into workplace design and workforce planning. There is a need for more personalized workforce planning/ personalized models that consider individual cognitive abilities, preferences, and fatigue levels to optimise both individual and team performance.

- Research on cognitive abilities often relies on general tests, but real-world work tasks require more specific cognitive assessments (sharp brains) that take into account context, task type, and environmental conditions. Although cognitive load has been studied, there is a lack of consensus on how to measure CLAM effectively and consistently in real time in dynamic environments. The specific application of CLAM factors to manual assembly tasks remains under investigated, particularly how adaptive mechanisms are employed when cognitive load increases due to complex tasks or high-paced environments. Combining wireless EEG data with other cognitive workload measures may provide richer insight into cognitive workload and performance, but it is still underexplored.
- Current ALBP models assume homogeneous workforces, but individuals vary in cognitive and physical capabilities. Research on how ALBP can be customised for individual worker profiles would enhance performance. Moreover, the weighted-sum method relies heavily on the assigned weights, and even small variations in these weights can significantly alter the objective function values. In the Pareto optimality approach, instead of assuming weighting for the three objectives, fitness is determined based on a non-dominated solution. This approach can be applied in the solution of ALBP considering ergonomic aspects.

2.13 Summary of Chapter 2

The major research topics considered in this research based on literature review are influence of cognitive aspects affecting human performance in manual assembly, worker allocation, cognitive abilities, CLAM factors, experimental studies using wireless EEG headset and assembly line balancing problem. There are multiple topics involved in a vast variety of various fields and identifying the keywords and relating to assembly line is done. This chapter has reviewed and investigated the major research topics identified. The experimental setup is inspired from the literature review and also the methods used for measuring cognitive load. The various methods to measure cognitive load are subjective methods, performance-based methods and physiological methods. Cognitive abilities were identified, CLAM factors were identified and then a framework is made for worker allocation with the experimental output and ALBP.

Ergonomic applications in assembly lines are very much limited in developing countries. Ergonomic assessments should be done in the manufacturing industry and the improvements should be made to obtain better productivity. Low-cost ergonomics solutions have to be designed for workers, particularly in developing countries. Future research can also focus on the comparison of ergo friendly work station with a traditional one and their output must be compared and analysed. The literature gaps identified from the review of literature helps to identify the research questions and research problem. The problem description, research problem, research questions and solution methodology are discussed in the next chapter. This review could guide the development of ergonomic friendly workstations for the workers which in turn improve their efficiency and reduce reworks due to human errors.

Chapter 3

Problem Description and Research Methodology

3.1 Introduction

The literature review presented in the previous chapter confirmed that cognitive aspects affecting human performance in manual assembly, workforce planning, cognitive abilities, CLAM factors, experimental studies based on wireless EEG headset and ALBP are the major keywords used in this research. These topics are considered to identify the worker's performance and cognitive abilities to allocate the workers based on their individual abilities. Ergonomic interventions will help the workers to improve their efficiency and employers to achieve productivity with minimal errors. Based on the findings from the literature review, section 3.2 formally defines the research problem. The solution approaches adopted are outlined in detail in section 3.4. This chapter is concluded in section 3.5. Figure 3.1 depicts an overview of thesis linking how the different chapters are related to the seven Research Questions (RQ), experimental studies and methods used.

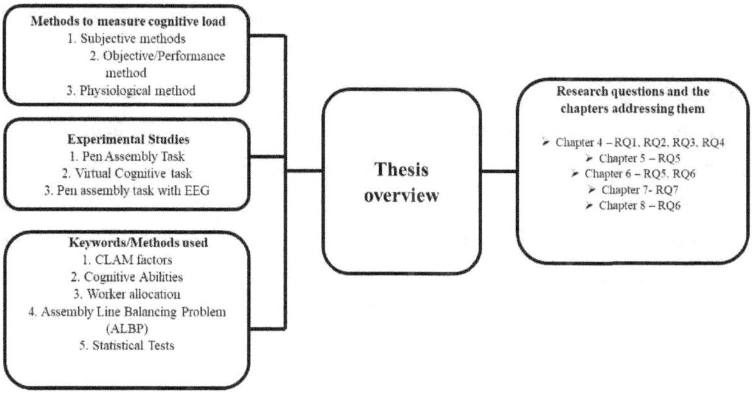

Figure 3.1 Thesis overview

3.2 Research Problem

An increase in customer demands and the advent of new technologies resulted in a multiple variety of products with a varying number of parts. The level of product variation leads is higher and this leads to an increased cognitive load for the worker. This demands on proper information and material presentation to be provided to the worker to fulfil the assembly task. The cognitive load for each worker differs as their cognitive abilities are different. There is need for an assessing the workstation of manual assembly workers and identifying their performance. In the existing literature a proper methodology for work allocation is not reported. The research problem is explained in the following sub-sections.

3.2.1 Influence of cognitive aspects of information and material presentation in a manual assembly environment

The study found that worker contributions are crucial to overall performance in most manufacturing units. In the laboratory, a pen recycling unit from southern India, an initiative of a non-profit organization aimed at reducing plastic waste and emphasizing the importance of recycling is replicated as a manual assembly unit. The cognitive load of the workers will be on higher with the increasing variety of products in a manual assembly unit. CLAM factors are developed for identifying and reducing the possible cognitive load among assembly personnel in a manufacturing context [3]. The CLAM factors are identified based on the pen assembly unit. An investigation on the influence of cognitive aspects namely information presentation and material presentation on worker performance in a manual assembly line [20 - 22] is considered and worker allocation [29] is done. An experimental study involved fifteen workers who were presented with eight different scenarios with six varieties of pens to be assembled. As a performance-based measure the number of pens assembled and the errors are tabulated. Further, a subjective measurement using NASA-TLX [11] is done. The study also identifies the individual worker performance and the importance of considering cognitive aspects in a manual assembly. The worker allocation methodology proposed can be used by similar manual assembly workstations.

3.2.2 Identifying the cognitive abilities using virtual cognitive test

This study highlights the significance of cognitive skills in carrying out a manual assembly task. The significance of these abilities in an experimental study has not been widely explored. Eight cognitive abilities [6] such as perception, attention, memory, motor skills, language, visual and spatial processing, executive functions and learning are considered in the virtual cognitive task. For testing each cognitive ability, brain teasers (sharpbrains) are used, and the subjects are rated accordingly. The weights of these cognitive abilities are varied, and thus the influence of these abilities on the total score of the subject is analyzed using statistical analysis. It is inferred that the rankings of cognitive abilities remain unchanged even when different weights are assigned to the cognitive ability scores. Cognitive neuroscientists agree that ongoing progress in understanding human cognitive skills would require not only new data acquisition, but also data synthesis and integration through studies and laboratories. Efforts for an increased focus on cognitive skills and their effective way of identification of tasks and experimentation using cognitive concepts [60] contributed to the framework of the study.

3.2.3 Worker allocation using personalized models from cognitive abilities

This study is a combination of the performances of subjects in pen assembly task and virtual cognitive task to build personalized models based on the cognitive abilities required to perform the pen assembly task. Statistical methods namely, Poisson regression and omnibus test are utilized. It is observed that visual perception, memory, visual and spatial processing, executive functions and learning are the major cognitive abilities required to perform the pen assembly task. The weightage of each cognitive ability for the best performance is identified and based on that worker allocation [29] is done. Out of the thirty subjects only six of them were eligible to work in the assembly task considered to provide better productivity. In practical situations this won't be possible as the employers will have to incur losses, a novel methodology is proposed wherein all the workers can be allocated based on their cognitive abilities. With effective allocation only one worker requires training which further reduces the cost.

3.2.4 Investigation of pen assembly worker performance using wireless EEG headset

A wireless EEG headset [56] is used to validate the pen assembly task. The pen assembly task is repeated with the wireless EEG headset. An increase in cognitive load is measured by an increase in brain activity, and the variation is identified for each scenario. The experiment with the wireless EEG headset - Emotiv Epoc X revealed that the observed differences in band power among the different scenarios are sensitive to changes in cognitive load. When the amplitude of these frequency bands increases, it suggests a higher cognitive workload, indicating that tasks in certain scenarios were more demanding and complex for the subjects.

3.2.5 Effect of ergonomic aspects on single and mixed model assembly-line balancing problems

This research incorporates ergonomic considerations into an assembly-line balancing problem [63] which enhances productivity and minimises ergonomic concerns. The assembly process, characterized by repetitive motions and handling numerous components, can lead to worker overload. This study investigates a multi-objective ALBP aimed at minimizing the number of workstations, overall skill level required, and variance in workers' energy expenditure across workstations. To address the ALBP while considering the ergonomic aspects, this study proposes an approach based on the Non-Dominated Sorting Genetic algorithm II (NSGA-II) and Multi-Objective Simulated Annealing (MOSA) using Pareto optimality. In the existing literature, researchers typically employ a weighted-sum approach [9] to address multi-objective problems in line balancing. However, the drawback of the weighted-sum method is its reliance on the assigned weights, where even a small variation can significantly alter the objective function values. A comparative analysis of NSGA-II and MOSA was conducted in single- and mixed model production scenarios to identify the dominant algorithm. The computational analysis compared the runtime performance of MOSA and that of NSGA-II.

3.3 Research topics and research questions

The major research topics related manual assembly task based on the literature review identified are as follows:

- Influence of cognitive aspects affecting human performance in manual assembly
- Worker allocation
- Cognitive abilities of subjects
- CLAM factors
- Experimental studies based on wireless EEG headset
- Assembly Line Balancing problem (ALBP)

The research questions formulated based on the above listed research topics are listed below. The research questions addressed in the present research are as follows:

RQ1: What are the various factors influencing human cognitive load?

RQ2: How can the cognitive load be varied?

RQ3: Is it possible to perform a cognitive load assessment for manual assembly?

RQ4: Do the information presentation and material presentation in a workstation influence the performance of an individual?

RQ5: What are the cognitive abilities/brain functions and how it influences the performance of experimental tasks done in manual assemblies?

RQ6: Is it feasible to evaluate human cognitive performance to create a framework for worker allocation?

RQ7: How can physiological measurements like EEG be applied to measure cognitive load?

3.4 Research methodology

This section discusses the research methods adopted for the experimental studies related to this work.

3.4.1 Experimental work using a Pen Assembly Task

The literature review identified several factors that affect the performance of worker at manual assembly workstations. Out of these factors four were considered for further investigation in this work, as they are involved in worker performance and worker allocation. The four factors are:

- Product varieties
- Workstation layout (material presentation)
- Worker instructions (information presentation)
- Operator's own way of working

The above factors identified, can be studied using the CLAM factors and hence the CLAM factors to be considered for the pen assembly task are identified. The cognitive aspects information and material presentation used in the study are mapped to the CLAM factors. The pen assembly task is done by the subjects in a laboratory environment where the experimental setup is similar to the workstations in pen recycling unit. There are eight scenarios and the number of pens assembled without errors is considered and a Kruskal Wallis non-parametric test is done for three sets of hypotheses. Cognitive Load Theory is used and the intrinsic, extraneous and germane load for all the fifteen subjects are calculated. NASA- TLX [11] is used as a subjective method to measure cognitive load. The number of pens assembled correctly based on the pen types which are assembled is noted by utilizing the concept of cognitive penetration and aesthetics Cognitive penetration refers to the idea that an individual's perceptual experiences can be influenced by their beliefs, expectations, and other cognitive states, while aesthetics explores the nature of beauty, art, and taste, and how they relate to human experience. Burnston [41] claimed that the cognitive penetration and aesthetic engagement has a secure relationship. A logistic regression model has been developed using the data obtained. This proposed model classifies the workers in terms of whether they will assemble the pen without error or not. The worker allocation is done based on the workstations in the pen recycling unit and a methodology was framed.

3.4.2 Experimental work using virtual cognitive task

This experimental study is done in a virtual mode to test different cognitive abilities. The tasks are given based on the brain blog research websites which integrate brain health into training, leadership programme. A cognitive test using all eight cognitive abilities was taken from a brain blog website. Each subject performed different tasks/ responded to the questions asked respective to the eight cognitive abilities and their skills involved [6]. The differences in the scores of the two matched groups similar to the t-test for two related samples are done using Wilcoxon Signed Ranks Test. Equal weightages score is out of 80 and unequal weightages score is out of 73. A hypothesis is defined to suggest that whether the rankings will be the same or not when the cognitive ability scores are given different weights.

3.4.3 Worker allocation using personalised models from cognitive abilities

The results of pen assembly task are compared to virtual cognitive task to identify the cognitive abilities required to perform the pen assembly task. Personalized models with different combinations of cognitive abilities are considered. Worker allocation of thirty subjects is done based on a methodology framed. Four different CLAM factors are used and the variation of these CLAM factors based on the complexity of four tasks are identified. Hypothesis testing with six sets of hypotheses which are the combinations of different tasks are considered. Poisson regression is used to model count variables. It has a number of extensions useful for count models. The statistical test used is Omnibus Test (Test that all of the estimated coefficients are equal to zero–a test of the model as a whole. From the p-value, we can see that the model is statistically significant.) and the level of significance is 5%. The independent variables are the eight cognitive abilities and the dependent variables are the total number of pens assembled correctly. The parameter estimates are also considered to identify the values of each cognitive abilities. This includes the regression coefficients for each of the variables along with robust standard errors, p-values and 95% confidence intervals for the coefficients. From the p-value, we can see that the model is statistically significant. The indicator variable [B value] is the expected difference in log count between a certain group and the reference group. Regression coefficients indicates the expected difference in log count between a group and its reference group. Highest regression coefficient (B) is selected as the best combination while considering the groups of three cognitive abilities. The significant cognitive abilities and their scores required to complete pen assembly task is obtained. A worker allocation methodology is framed for thirty subjects.

3.4.4 CLAM FACTORS

Cognitive Load Assessment in Manufacturing method [3] is a useful framework and method for considering cognitive aspects in a manual assembly. CLAM was developed for identifying and reducing the cognitive load of manual assembly workers. Identifying relevant issues at the assembly workstations can lead, for instance, to saved time and resources on the shop floor. This framework also presents a connection between worker cognitive load and manual assembly.

Table 3.1 CLAM factors and their description

CLAM factors	Description
Saturation	The amount of work that is planned on a workstation related to the particular balance of the assembly task
Variant flora	A product or process variation from the most common type of product (volume). Mostly an issue in mixed mode assembly flow Strongly connection to cognitive workload
Level of difficulty	A subjective estimation about the required physical and cognitive effort to perform a task Heavily tied to required time of necessary training and skills needed to perform task independently

CLAM factors	Description
Production awareness	Refers to how much focused /active attention that must be applied to the task and the level of "production awareness" that the worker has to muster
Difficulty of tool use	Refers to both the amount of tool use required but also the estimated complexity of the tool use. Including all tool use. even special or non- standard tools
Number of tools	The number of tools used during a normal assembly task, including special and non-standard
Mapping of workstation	Refers to how well the workstation design maps with the assembly sequence Tools and parts that are used together should be placed together and in the correct order.
Parts identification	The identification syntax used at the workstation such as components numbers and material racks or kitting
Quality of instruction	Refers to on a general level how visible and readable the instructions are used to gather information about the work

Table 3.1 (Contd.)

CLAM factors	Description
Information cost	Refers to how much physical and cognitive effort is required to utilize the information
Poka-yoke	Using poka-yoke solutions or constrains to reduce assembly errors. Including designing the task and/or product in order to prevent assembly errors

As seen in Table 3.1, the CLAM factors will be a valuable input in correlating it with the manual assembly tasks and cognitive load. The selected factors from CLAM which could be connected to information presentation and material presentation are shown in Table 3.2.

Table 3.2 CLAM factors used for Pen assembly task

Factors used in the exploration studies	Factors selected from CLAM
Material presentation	Mapping of Workstation
	Parts identification
Information presentation	Quality of instructions
	Information cost
Component variation	Variant flora

The CLAM factors used for varying the task complexity are:

1. Variant flora (Increases from task 1 to task 4)
2. Level of difficulty (Decreases from task 1 to task 4)
3. Production awareness (Increases from task 1 to task 4)
4. Quality of instructions (Increases from task 1 to task 4)

3.4.5 Methods to measure cognitive load

The tools used to measure a particular type of cognitive load can be divided into three main categories: subjective measures, performance-based measures, and psychophysiological measures.

1. Subjective methods: Self-report questionnaires or rating scales.
2. Performance-based methods: Evaluating outcomes, such as error rates, reaction times, and task completion times
3. Physiological methods: Heart rate variability and Electroencephalography

3.4.5.1 Subjective method

Subjective ratings are simple and easy to obtain but subject to individual biases, such as social desirability and adaptation or habituation to task demands. There two subjective methods in used in this research are NASA TLX and ISA. The pen assembly task is done with and without using wireless EEG headset and in both experimental studies, NASA-TLX is used. ISA is used for rating task complexity.

After finishing the pen assembly task in all eight scenarios, the subjects rated perceived workload on six different scales as per NASA- TLX: mental, physical, temporal, performance, effort, and frustration; the scales set from 0 to 100. The factors were weighted about one another to show the most significant aspects of the workload, with more importance given to mental demand. Excel macros is available to put the scores and to obtain the final index score. ISA is a technique that has been developed as a measure of workload to provide immediate subjective ratings while performing the task. The ISA is a self-rating method of the current mental workload during a task on a scale of 1 (low) to 5 (high). It is normally queried every 2-5 minutes during the activity. Kirwan [46] used the following ISA scale to assess air traffic controllers (ATC) workload

3.4.5.2 Performance based method/Objective method

Evaluating outcomes, such as error rates, reaction times, and task completion times, to infer cognitive load, Comparing the same performance measures under different levels of task complexity, stress, or cognitive load can be used to measure cognitive load. In this research the subjects cognitive load is inferred from their performance based on the total number of pens assembled correctly.

3.4.5.3 Physiological method

The two physiological methods used in the experimental studies are Heart Rate Variability (HRV) and Electroencephalography (EEG). Emotiv Epoc X is wireless EEG headset used and polar heart rate monitor is used for measuring heart rate variability. As per Cranford et al. [50] heart rate can potentially be used to measure changes in cognitive load and is appropriate to consider the data through the lens of the hypotheses. A baseline measurement of the heart rate was taken based on the task complexity of four tasks with two sets of hypotheses. An electroencephalogram (EEG) is a recording of brain activity which is a physiological method of cognitive load measurement. Wireless EEG headset can be used to determine cognitive workload by capturing the electrical activity in the brain based on frequency bands. The cognitive load of different scenarios of pen assembly task is captured by repeating the experiment with wireless EEG headset.

3.4.6 Optimization Approaches – Assembly Line Balancing Problem (ALBP)

The mathematical model that formed the basis of this study was obtained from the multi-objective optimization model developed by Dalle and Dini [9]. The energy expenditure for various micro movements in an assembly line was computed using the empirical relationships in Garg et al. [94]. Certain inputs are considered from the pen assembly task like the number of tasks, number of product types, number of workstations etc. In the present research, to solve an assembly line balancing problem considering ten tasks for single- and mixed model assembly, two multi-objective optimization solution approaches are proposed:

- Non-dominated sorting GA II (NSGA-II)
- Multi Objective Simulated Annealing (MOSA)

Contrary to the weighted solution approaches reported in the literature, a Pareto optimality-based approach was applied in the present study to obtain non-dominated solutions. This is the unique contribution of this study to the scientific literature on ALBP. In this study, the following two variants of ALB were considered:

1. Single-product ALB
2. Mixed model ALB

In both scenarios, ten tasks were involved in an assembly operation to be carried out at four workstations. Hence, these ten tasks are to be assigned among the available workers in the four workstations, such that the total required skills can be minimised. In addition, the variance in the total energy expenditure among the workstations should be minimised; that is, the task should be distributed according to the worker's physical capability.

3.5 Summary of Chapter 3

This chapter presents an overall problem description for the different stages of the research work. The ergonomic studies performed, framework and ALBP approach is provided in the chapter. The following five chapters in the thesis present the five studies that comprise the present research.

Chapter 4

Influence of cognitive aspects of information and material presentation in a manual assembly

4.1 Introduction

Modern manufacturing methods are evolving from mass production to mass customization, facilitating more frequent and consistent adaptations based on customer preferences, cutting-edge technology, and environmental considerations. This makes the system more complicated for allocation of a specific job to a worker based on their skills or allocation of a job to a newly joined worker, especially if the production system seeks to maintain a low cost while maintaining the quality of the goods [14]. A huge demand and complexity in the assembly process increases the stress among the workers. Grosse et al. [95] suggested a concept for integrating human factors into order-picking operations to boost productivity and improve workers' mental well-being.

The aesthetic appearance of the products attracts the workers during the assembly task, and this is known as cognitive penetration (CP). Hence factors involved in aesthetic perception are also to be considered while assessing the cognitive workload. Expertise that is mediated by instructions and learning may shape aesthetic perception. As stated by Burnston [41], the greater the understanding based on high-level and categorical learning, the more likely cognitive penetration occurs. According to the early formulation of Stokes [96], an association between cognition and perception exists. Cognitive load theory helps to identify the mental workload of the workers assigned. Galy et al. [4] reported the impact on cognitive load because of an additive relationship between intrinsic, extraneous, and germane load. In a workspace, if the health and protection of employees at work are crucial for productivity, the analysis of cognitive load and methods to reduce cognitive load are important. Hence, workers with a high cognitive load must be provided with additional material and information presentation.

In a mixed assembly mode with different variants of the product, Brolin [97] published a comparative study of information presentation and material presentation. In this study, the material presentation was varied between an unstructured package and a standard package; moreover, information presentation was compared between textual and number instructions with image instructions. When using images instead of text and numbers, assembly times and performance ratings were reported to be lower. However, in this work, the different combinations of structured and unstructured material presentations with textual and image instructions are considered for experimentation. This aspect is unique to this study. With the rising complexity and automation of assembly lines, Belletier et al. [98] identified the need for cognitive assistants, pushing human operators to apply different cognitive processes. Cognitive assistants assist human operators in analyzing, integrating, and maintaining large amounts of information. The information presentation used in this study can contribute to the development of cognitive assistants and optimize efficiency.

NASA Task Load Index (NASA-TLX) is a widely used subjective measurement method for determining performance aspects by rating perceived workloads. Emotional, physical, and temporal demands, dissatisfaction, commitment, and performance are the six dimensions used to operationalize the workload. Individual worker performance is known to be influenced by the cognitive aspects of workstation design. Worker performance variability is assumed to be largely insignificant in current workstation design methods. This study aims to identify workers' varying performance when different combinations of material and information are used in a workstation. In the allocation of workers to each work station according to their skills, it will be beneficial if it can be predicted whether the worker will assemble the pen without errors in a given scenario. In this direction, a logistic regression classifier model is developed in this work.

The attempts to achieve sustainability and the management of long-term environmental problems such as plastic waste recycling are important. Plastic recycling is given prime importance around the globe, being aware of the hazards caused by these pollutants. Recycling and reuse of products are different; the former is the reprocessing of a product into a new raw material from which a new product is produced, and reuse lengthens the life of an item. Compared to recycling, reusable goods have a decreased environmental impact, but manufacturers, for the most part, focus on techniques including light-weighting, recyclability and eco-labelling [99]. Actions that facilitate the reuse of products may improve sustainable behaviours [100]. In view of enhancing sustainable behaviours, this study focuses on establishing a manual assembly with improved productivity for a pen recycling unit, wherein the workers are assigned the task based on their performances.

The pen assembly task is an experimental study mimicked in the lab environment with subjects which is an analogy to the workers in the industrial environment and a methodology is created to allocate the workers based on their performance. This study shows that as the complexity and requirements of assembly operations in manufacturing increase, assembly workers' workload increases, leading to errors and a decline in productivity. New worker recruits also cause hassle because it becomes complicated to identify their skills and allocate them according to their skills. Previous research has shown the complexities an assembly worker faces in the work environment, but a solution based on the work environment with a case study is absent. This is the motivation for the present study.

4.1.1 Cognitive perspective on manual assembly tasks

This research focuses on worker allocation, considering the following two factors of cognitive aspects that affect human performance:

(1) Material presentation and

(2) Information presentation.

A mixed design approach of Creswell et al., [101] is used in this study which includes time, the number of pens assembled, and the success rate in the following eight scenarios of the experiment. Two types of kits have been described in previous research by Hanson and Brolin [20] as unstructured and structured kits, which can be used as a type of material presentation. The kit concepts motivated this study to define the factorial experiment design with various information presentation methods. This then constituted the four levels within the material presentation factor as follows:

- Intermixed parts
- Grouped parts
- Separated parts in different boxes
- Part numbered box

It was previously investigated whether various types of instructions had an impact on worker in a manual assembly. The recommendation to use text instructions and instructions using numerical, for information presentation in manual assembly, as well as photographs are based on previous studies by [20-22]. In this study the two types of information presentation used are as follows:

- Instructions
- Photograph

In essence, this research identifies the combined effect of information and material presentation on the individuals for a task. The combined effects identify the performance of each worker and further helps in assigning a worker to each workstation according to their cognitive skills. This study investigates the cognitive load theory and cognitive penetration. This study also explores how the performance of workers differ according to the information (instructions and photograph) and material presentation (intermixed parts, grouped parts, separated parts in different boxes, part numbered box).

Researchers suggest that photographs as a means of information presentation are more effective, but if it is considered through cognitive aspects, it can be observed that there needs further investigation on this criterion of information presentation. One of the hypotheses in this study states that while considering the performance between the part numbered box and separated parts in different boxes, there is a significant difference as per the analysis of results, the null hypothesis is rejected. In case of human performance while using photograph and the text instructions, there is a significant difference, and the null hypothesis is rejected. Thus, this output from the hypothesis suggests that the worker performance is varied according to the material and information presentation combinations. A logistic regression model is developed with the data collected. The model is tested with a part of the data collected and the findings from the model can support the allocation of workers to the proposed assembly stations for a pen recycling unit.

The contributions of this work can be enumerated as follows:

1. Investigate the combined influence of material and information presentation on the performance of workers according to the factorial design in a pen-recycling unit.
2. Development of a novel approach for worker allocation in pen-recycling unit considering the cognitive aspects obtained from the preliminary study.
3. Application of a classifier model to predict the performance of workers.

4.2 Environment and proposed method

The experiment was conducted in a small-scale pen recycling unit, which is an initiative of a non-profit organization located in southern India. The unit collects six variants of pens for recycling from nearby colleges, schools and other institutions. All the pens collected were difficult to sort and prepare for reuse. The main difficulty was sorting the pens as some were damaged, some parts were damaged, and the refill of the pen also needed to be restored. Hence, from the heap of used pens collected, all the pens were disassembled, and a quality check was done, and then further it was assembled. These collected pens were reused and sold for half the price. The idea was to frame an assembly line and to set up the whole process, from separating the parts from the heap of used pens to assembling a recycled pen with a new refill. The workstations were identified as five and only could accommodate fifteen workers as the unit has a lower capital.

In view of the requirement, a factorial design was framed with eight scenarios that match the tasks to be done in each workstation of the pen-recycling unit. This technique helps to identify the best performers in each workstation and the analogy of the scenarios and workstations helps to allocate them. All six types of pens are considered in the experimental study, which involves cognitive aspects. The subsequent sub-sections provide further details about the experiment.

The study is conducted in two phases, where half an hour break is provided for the workers just to relax between two phases. The first phase includes scenarios 1 to 4, and the second phase includes scenarios 5 to 8. Each worker is given a time limit of 10 minutes. Further, a quality check of the parts assembled is done after the experiment, and the number of pens assembled is noted. The information and material presentation have been modified for the proposed experiment, but the methodology is different. Participants, task design, and methodology of this study are discussed in the following subsections.

4.3 Participants

A total of 15 participants aged between 20 and 30 years, with a mean age of 21.93 years and a standard deviation of 3.23 years are utilised for this experiment. Given an effect size of 0.68, generated from a prior study [97] utilizing the identical information and material presentation combinations, a sample size of 15 is estimated using G-Power for a power level of 0.96. Two significant interaction effects with a considerable impact size between information presentation material presentation are considered. This means that when information presentation and material presentation in the factorial design were combined, certain combinations were better than the others. The selection process ensured that the workers had sufficient awareness about the different parts of the pen. It is ensured that all inclusion requirements are fulfilled by the workers, such as no stress, tension, nobody ailments in the last 24 hours. The workers were informed about the objectives of the experiment, tasks to be completed. A duration of ten minutes was provided for each worker to complete each task in a scenario, and the time was monitored using a stopwatch. The number of pens assembled is noted and each pen is disassembled and checked for errors in each scenario.

Fifteen workers are considered based on the requirement of workers in each workstation of the pen recycling unit. A total of five workstations in this pen recycling unit requires three workers in each workstation, so depending upon the production demand in the pen recycling unit, the number of workers is selected. Depending on the requirements in a real-time assembly line, if the number of workers is increased in the experimental study, then the number of workers assigned to a particular workstation will be more than the number of eligible workers required. This increases the complexity in the worker allocation process, workstations lack eligible workers, and thereby productivity is affected. The employer must reconsider the number of workers for the study or give appropriate training for the workers to work in an assigned workstation.

4.4 Task Design

The tasks performed in the eight scenarios of the experimental study are analogous to the tasks carried out at workstations. There are eight scenarios with four workers at a time in the experimental setup. The time taken and errors made are not revealed to the workers as this may influence their remaining task. The six varieties of pens and their parts are introduced, the different types of instructions to provide information of what they will assemble, and an exploded view of each pen to be assembled are displayed. The workers can select any variety of pens to assemble, but the six types of pens have to be completed before the assembly of the next set of pens. The material presentation involved the following four levels: (i) intermixed parts, (ii) grouped parts, (iii) separated parts in different boxes, and (iv) part numbered boxes. The intermixed parts contain six varieties of the pen in a single box with all the parts disassembled and put together, as shown in Figure 4.1.

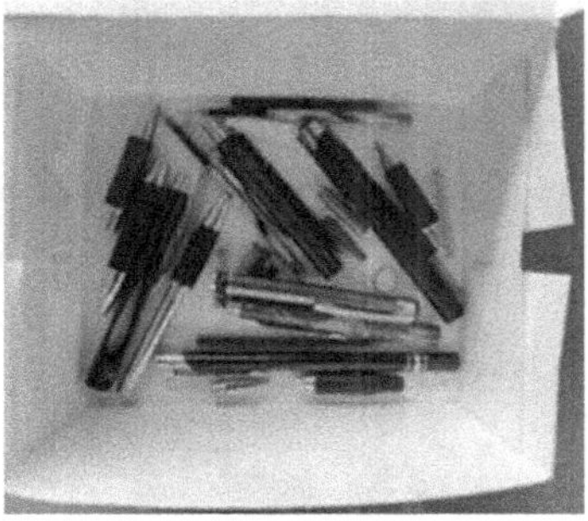

Figure 4.1 Intermixed parts

The grouped parts consist of twelve pens with two sets. In this type, the barrel, tips, refill, and other parts of all the pens are grouped according to their part family, as shown in Figure 4.2.

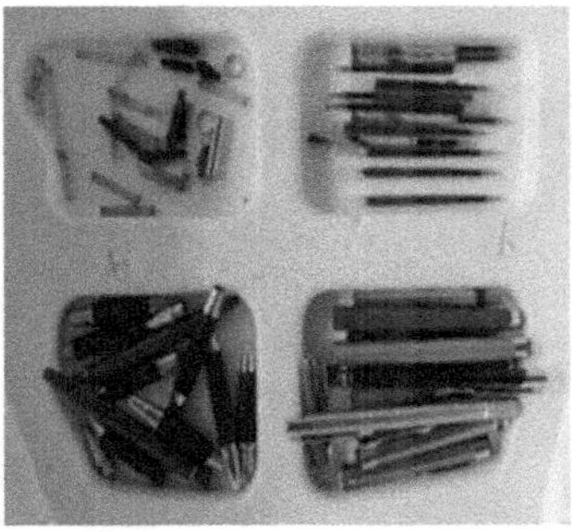

Figure 4.2 Grouped parts

The third level separated parts in different boxes have a total of thirty pens disassembled, having five sets of six varieties. In this, each part of each pen is in different boxes, and the workers have to pick and assemble according to the information presentation. The variation in parts of different pens helps to find out the cognitive abilities of this task. A detailed top view of the setup is shown in Figure 4.3.

Figure 4.3 Separated parts in different boxes

The part numbered box is similar to the separated box, but the box cover includes the part numbers by which the part family can be identified so that the pen can be assembled in an easier way. As shown in Figure 4.4, the part numbers are arranged in random order, and the workers have to identify the part numbers and then assemble them.

Figure 4.4 Part numbered box

The information presentation included photographs and text instructions. The instructions include detailed information about parts of each pen with which assembly of the different parts will be made easier for the worker. The instructions for part numbers with the grouping of parts for each pen is provided by grouping the parts to a family and provided with number codes. The scenarios followed the order as given in Table 4.1, which shows the entire nature of the experiment, including the level of factors.

Table 4.1 Factorial Experiment Design

Scenario No	Material Presentation	Information Presentation
1	Intermixed parts	Instructions
2	Grouped parts	Instructions
3	Intermixed parts	Photograph
4	Grouped parts	Photograph
5	Separated parts in different boxes	Instructions
6	Part numbered box	Instructions
7	Separated parts in different boxes	Photograph
8	Part numbered box	Photograph

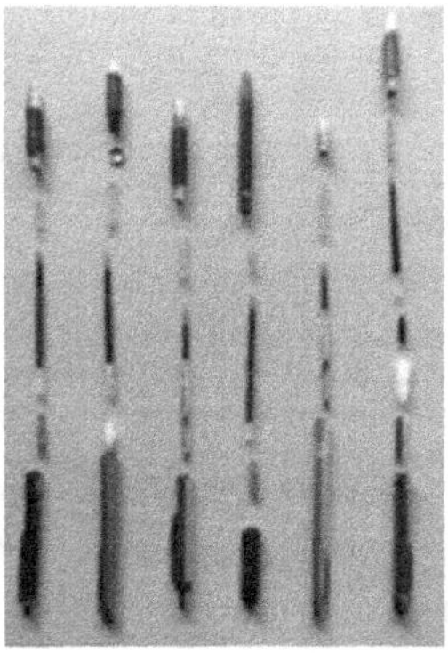

Figure 4.5 Exploded view of six varieties of pen

4.5 Experimental Methodology

A relevant framework, CLAM, was derived from the usability approach after being modified by several researchers. This method assesses the cognitive factors related to manual assembly. CLAM method helps to educate and assist manufacturing companies in reducing cognitive load on the shop floor. There are eleven CLAM factors and from these five of the factors are considered, according to the experimental set-up and the workstation for the pen recycling unit. Mapping of workstations and part identification from CLAM is analogical to the material presentation. Information presentation includes quality of instructions and information cost. The need for six varieties of pen considered in this experiment is to consider the variant flora factor of CLAM. The experimental study with CLAM factors contributed to the cognitive aspects is to be taken care of in the workstations of the pen recycling unit.

Some of the highlights of the methods adopted in this study are as follows:

- Identify the various CLAM factors for the pen recycling unit
- Develop a factorial experiment design for a combination of information and material presentation, with the factors identified from CLAM. The study is conducted with fifteen workers as per the experimental design. The data collected from the study is used to allocate the workers on a pen recycling unit.
- The objective measurements are based on the errors the workers make while doing the tasks in each scenario. The number of pens assembled and the time for assembling a certain number of pens according to each scenario is noted. The performance is measured based on the relative performance of workers in different combinations of information and material presentation.
- The subjective measurements are done using the NASA-TLX assessment method. Cognitive load theory, cognitive penetration and aesthetics concepts are also included in this study.
- A logistic regression model which classifies the workers in terms of whether they will assemble the pen without error or not is also included.
- With the data collected from the experimental study, the workforce planning for the pen recycling unit is done based on the requirements in each workstation.

The experiment involves three steps, as shown in Figure 4.6. The workers are briefed about the experiment in the first step. In the second step, workers are given different combinations of information and material presentation. As the workers carry out the task, they are monitored and the performance of each worker is noted. At the end for subjective assessment, a questionnaire survey is conducted to measure the perceived cognitive load for the task.

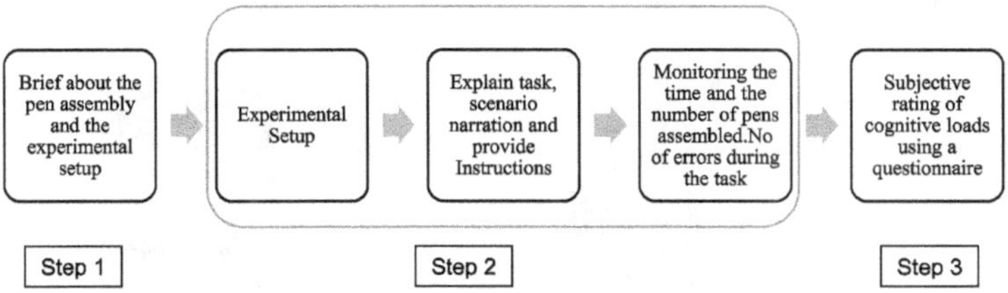

Figure 4.6 Experimental Methodology

The scope of the analysis comes from many different sources, which is one of the key starting points of this study, as well as other field research. An assembler, for example, gains knowledge not only from the instructions but also from the work environment and co-workers of the assembly. This study, therefore, explores how the assembly operator is influenced by multiple factors at a workstation. There are six types of pens in this sample, as shown in Figure 4.7, with differing numbers of parts and are to be assembled in eight different materials. Two different information presentation combinations are provided.

Figure 4.7 Illustration of six varieties of pen

This experiment is to analyse the performance of workers in a pen assembly task and to allocate workers based on the information presentation perspective, the number of parts assembled and the number of errors in each stage. Workers will be in an isolated room to execute the tasks with eight scenarios, as shown in Figure 4.8.

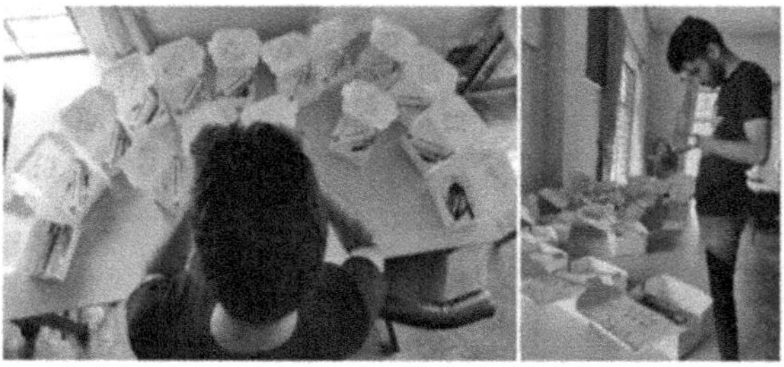

Figure 4.8 Experimental setup.

The objective measurements adopted in the present work are as follows:

- **Time** – The time required for each worker to assemble six varieties of pen and this time is fixed for all workers.
- **Error** – The number of errors in a pen which is assembled. In this experimental study, two types of errors were recorded: (1) picking the wrong part while assembling a unit correctly and/or (2) picking the correct part but assembling it incorrectly. The parts of the pens assembled will be checked. The errors are characterized based on either the difference in spring, refill, adding an extra part to a pen, and missing of small parts like O-rings, tips etc.

The majority of workers stated the experiment was interesting and the task seemed to be easy. Photographs were the best choice at first, but after a while, they were no longer relevant, except for the type numbers beneath the images. Instead, the individuals assembled the parts instinctively, which resulted in assembly problems as the part variants changed. The NASA-TLX method was used to perform the subjective measurement.

4.6 Hypotheses

The three hypotheses used for this factorial experiment (HA, HB and HC) concern the different levels of the factors: presentation of material (HA and HB), presentation of information (HC).

H0A: No significant difference between the performance when using grouped parts and the performance when using intermixed parts.

H1A: Significant difference is observed between the performance when using grouped parts and the performance when using intermixed parts.

H0B: No significant difference between the performance when using part numbered box and the performance when using separated parts in different boxes.

H1B: Significant difference is observed between the performance when using part numbered box and the performance when using separated parts in different boxes.

H0C: No significant difference between the performance when using a photograph and when using instructions.

H1C: Significant difference between performance when using a photograph and when using instructions.

4.7 Results and Analysis

This study considers 5 out of 11 CLAM factors by identifying the process in pen recycling unit. A comparison of the results obtained from Kruskal Wallis test from the pilot study and the present study is similar which verifies the validity of the pilot study and performance of the workers based on the cognitive aspects.

The human performance measurements are important so as to get a detailed idea about the cognitive aspects. The objective and subjective measurements contribute to the results in the experimental design which we can correlate with the corresponding workstations in the study. The pen recycling unit has 5 workstations and the whole study relates these workstations with the scenarios of experimental design. There is an analogy between scenarios 1,3, and workstation 1 and scenarios 2,4 and workstation 2. There is an analogy between scenarios 5,7, and workstation 4 and scenarios 6,8 and workstation 3. All the four sections discussed below contributes to the worker allocation.

4.7.1 Statistical analysis -Kruskal Wallis test

In order to test the normality of the data, the Kolmogorov-Smirnov goodness of fit test was performed. The results revealed that the data does not follow a normal distribution, and hence to test the hypotheses, a non-parametric test of Kruskal-Wallis at 95 percent confidence is done using a commercially available statistical package. With the results, it was identified that some of the main effects based on the hypothesis considered are significant in the analysis of results. One of the hypotheses states that while considering the performance between the part numbered box and separated parts in different boxes, there is a significant difference as per the analysis of results, the null hypothesis is rejected. In the case of performance between the use of the photograph and the text instructions, there is a significant difference, and the null hypothesis is rejected. Thus, this output from the hypothesis suggests that the worker performance is varied according to the material and information presentation combinations.

The results of the Kruskal-Wallis non-parametric test based on the hypothesis are as follows:

- In the material presentation, there is no significant difference in performance between grouped parts and intermixed parts ($p > 0.05$). The p-value of the Kruskal-Wallis Test is greater than 0.05, which fails to reject the null hypothesis (H0A).
- In material presentation, there is a significant difference in performance between part numbered box and separated parts in different boxes ($p < 0.05$). The p-value of the Kruskal -Wallis Test is less than 0.05 and the null hypothesis (H0B) is rejected.
- In information presentation the performance between the use of the photograph and the text instructions, there is a significant difference as per the analysis of results ($p < 0.05$) and the null hypothesis (H0C) is rejected.

The first four scenarios of material presentation are found to be similar and the worker performance does not vary much. In the case of the last four scenarios, when it comes to part numbered boxes and separated parts in different boxes the material presentation influences the performance, similar is the case with photographs and instructions. The combinations of information and material presentation are categorized into eight different scenarios as shown in Table 4.1.

Table 4.2 Experimental Results

Worker	Total	Max	Min	Scenario for Max	Scenario(s) for Min	Success Rate
1	21	7	0	Scenario 6	Scenario 1 & 5	0.1346
2	26	11	0	Scenario 8	Scenario 1,2,5,7	0.1667
3	43	13	0	Scenario 6	Scenario 2	0.2756
4	36	11	0	Scenario 8	Scenario 2,5	0.2307
5	28	13	0	Scenario 8	Scenario 2	0.1795
6	47	13	0	Scenario 8	Scenario 1	0.3013
7	26	8	0	Scenario 6	Scenario 2	0.1667
8	21	10	0	Scenario 8	Scenario 1,2	0.1346
9	26	10	0	Scenario 8	Scenario 1,2,5	0.1667
10	29	14	0	Scenario 8	Scenario 1,2,4,5	0.1859
11	36	14	0	Scenario 8	Scenario 2,3,5	0.2307
12	38	19	0	Scenario 8	Scenario 1,2,3,5,7	0.2435
13	22	6	1	Scenario 6	Scenario 1,2,5	0.1410
14	13	4	0	Scenario 6	Scenario 2,5	0.0833
15	30	12	0	Scenario 6	Scenario 1,2	0.1923

The experiment is conducted with the workers who are to be assigned to the workstations. The results reveal that workers have assembled a maximum number of pens in scenarios 6 and 8, i.e., a scenario where the workers are provided with part numbered boxes with instructions or photograph. However, there exist variations in the minimum number of pens assembled. This variation is due to the change in their skillset and is to be applied in the allocation of the workers. A detailed analysis considering the number of pens assembled by each worker is also done to identify which combinations yielded better results individually. The success rate of the tasks is very low because the workers have no prior experience with them. Even though the success rates are low, when considering the learning effect from the lower number of pens assembled in scenario 1 to a greater number of pens in scenario 8 provides an illustration of cognitive assessment usage. The success rates of the workers with the total number of pens assembled and the scenarios with maximum and minimum are provided in Table 4.2.

4.7.2 Human Performance Measurements

4.7.2.1 Objective measurements

The difference in information presentation (instructions and photograph) based on the number of pens assembled is given in Table 4.3.

Table 4.3 Number of pens assembled based on Instructions and Photograph

Workers	Instructions	Photograph
1	8	13
2	10	16
3	15	28
4	12	24
5	9	19
6	10	37
7	12	14
8	5	16
9	6	20
10	10	19
11	15	21
12	15	23
13	9	13
14	6	7
15	13	17

The total number of pens assembled using instructions is 155, and using photograph is 287. All the workers had done a better job when information presentation is in the form of photograph in the whole experiment.

The combination of material presentation using part numbered box and information presentation using photograph (in scenario 8) is resulting in the best output, which means that the combination of these types of presentations in the work scenario provides better productivity. Part numbered box with Instructions (in scenario 6) has a significant output as in scenario 8; the difference of performances in other scenarios with that in scenarios 6 and 8 justifies the effect of cognitive processing among the workers. The instructions with grouped parts (in scenario 2) showed the lowest number of assembled pens; this can be due to the difficulty in identifying the parts of individual pens. In scenario 2, the workers adopted mostly a trial and error method, by which they could not complete the assembly of pens within the stipulated time. Table 4.4 presents the combination of four types of material presentation and two types of information presentation in the eight scenarios considered in the present study. Table 4.5 tabulates the number of pens assembled in each scenario.

Table 4.4 Scenarios based on Instructions and Photograph

	Intermixed	Grouped	Separated parts in different boxes	Part Numbered Box
Instructions	Scenario 1	Scenario 2	Scenario 5	Scenario 6
Photograph	Scenario 3	Scenario 4	Scenario 7	Scenario 8

Table 4.5 Number of pens assembled based on Instructions and Photograph

	Intermixed	Grouped	Separated parts in different boxes	Part Numbered Box
Instructions	13	4	9	129
Photograph	32	54	53	148

Further, a detailed analysis of each scenario and individual performances of the workers are carried out. It is observed that in both the scenarios 6 and 8, which provided the best results, the workers 1,3,7,13,14, and 15, contributed to the best performance of assembling maximum number of pens using part numbered box with instructions. These results lead to the importance of material and information presentation provided to the workers for improving the performance. Workers may present high skills in assembly process but the workstation design should help in enhancing worker performance. Based on the results from this study for workstation design photographs as information presentation is better than instructions. Figures 4.9 – 4.12 depicts the number of pens assembled by workers in different scenarios.

Figure 4.9 Number of pens assembled Vs Workers for Scenarios 6 and 8

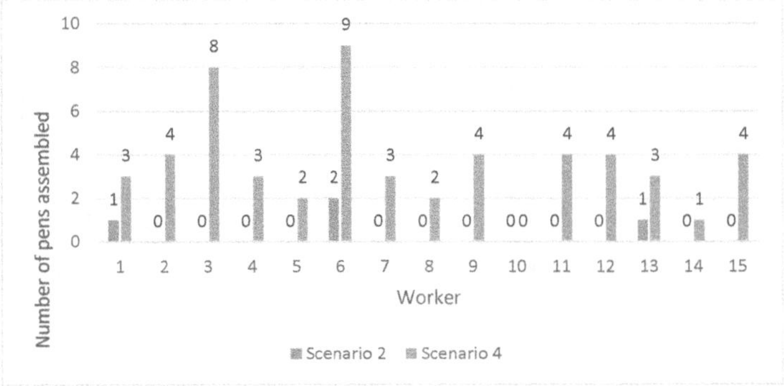

Figure 4.10 Number of pens assembled Vs Workers for Scenarios 2 and 4

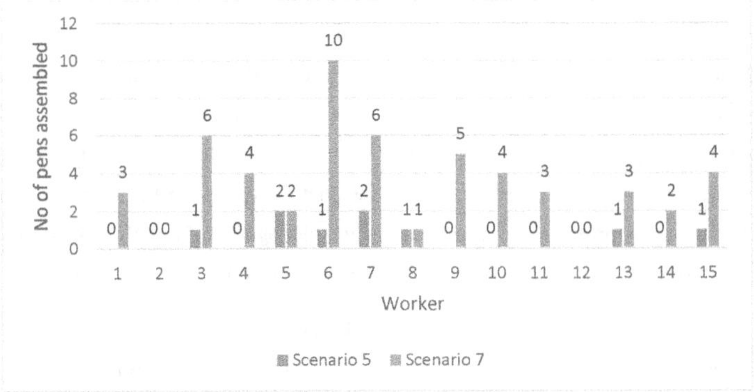

Figure 4.11 Number of pens assembled Vs Workers for Scenarios 5 and 7

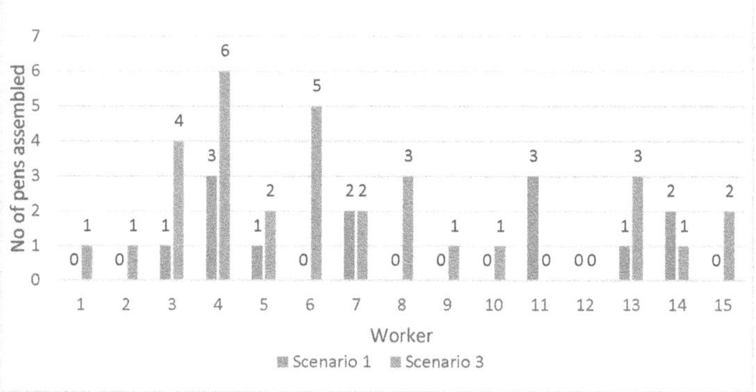

Figure 4.12 Number of pens assembled Vs Workers for Scenarios 1 and 3

It is observed that in scenario 2 (Grouped parts with instructions), least number of pens are assembled. A significant difference exists between the performance of workers in scenarios 2 and 4. Therefore, it is clearly evident that the performance of workers provided with Grouped parts and photograph is better than that of workers with grouped parts and instructions. It can be noted that workers 2 and 12, could not assemble a single pen by providing both instructions and photographs. Thus, there was no output in each scenario. An equal number of pens are assembled by workers 5 and 8.

Considering scenario 5 and 7 (Separated parts in different boxes with instructions and photograph respectively), worker 5 has equal number of pens assembled, which is to be considered during the final allotment as there can be variations in the performance if the photograph instruction method is provided. Workers 2 and 12 could not assemble any of the pens in these both scenarios. All the remaining workers gave a better output in separated parts in different boxes with photograph.

In Scenario 1 and 3 (Intermixed parts with instructions and Photograph respectively), worker 11 and 14 had assembled very few numbers of pens, that too with instructions. Therefore, it is evident that if these workers were allotted to the scenario with photograph there are chances for their output to be zero. Worker 12 also could not assemble any pens and this can be due to the fact the there is no previous information about the task to be done, but as the task progresses, the worker could assemble a good number of pens.

This section thus confirms that there will be differences in performances of the workers in different scenarios when considered individually and this has to be further studied and analyzed for better performance and output in industry environments. It is evident that depending on the worker, the material and information presentation has to be varied. This comparison also contributes to the allocation of the workers in each station.

4.7.2.2 Subjective measurements

- **Subjective Measurement for Cognitive Load**

A subjective measurement of cognitive load is performed at the end of the experiment. A questionnaire survey is conducted for rating the difficulty of tasks. Leppink et al. [102] has proposed a subjective rating technique for measuring all the three types of cognitive load. Adopting a similar approach, this study measured the cognitive loads. It is observed that the intrinsic load is highest for worker 8 and the lowest for worker 12. The intrinsic load of workers is depicted in Figure 4.13.

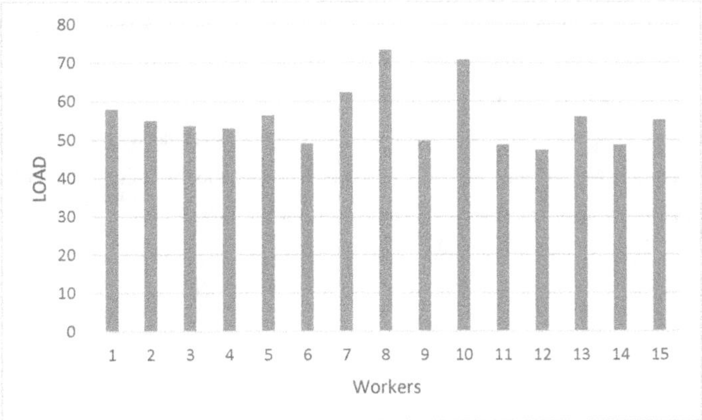

Figure 4.13 Intrinsic load of workers

The extraneous load is less than the maximum intrinsic load. The results indicate that worker 7 and worker 8 has the least extraneous load but they are not the ones with the highest number of pens assembled. The performance can be identified from this data as per the graph shown in Figure 4.14.

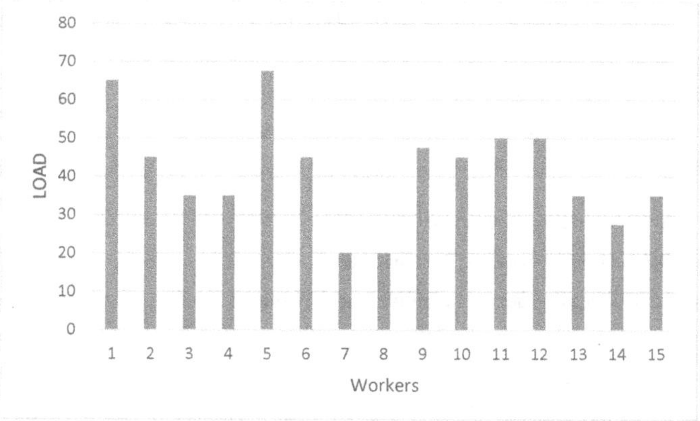

Figure 4.14 Extraneous load of workers

Germane load is mostly dependent on the effort, mental demand and the frustration level. The germane load is highest for worker 10 and this is the highest value among all the three loads discussed here. In this task, initially, the cognitive load is higher as they are not familiar with the assembly line. The scenarios from one to eight made them learn and at the end of the eighth task, the cognitive load had decreased. It is evident that the number of pens assembled based on a scenario by individual workers has increased. Figure 4.15 presents the germane load of workers.

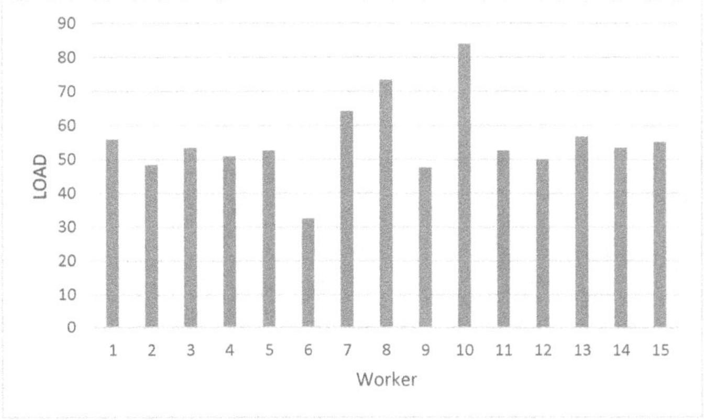

Figure 4.15 Germane load of workers

- **NASA Task Load Index (NASA-TLX)**

NASA-TLX is an assessment method used for subjective measurement of cognitive load in this study. NASA-TLX is a workload evaluation tool used to investigate the expected workload. After finishing the task of assembling the pen in all eight scenarios, fifteen workers rated perceived workload on six different scales which are mental, physical, temporal, performance, effort, and frustration and the scales are set from 0 to 100. The factors were weighted about one another to show the most significant aspects of the workload, with more importance given to mental demand.

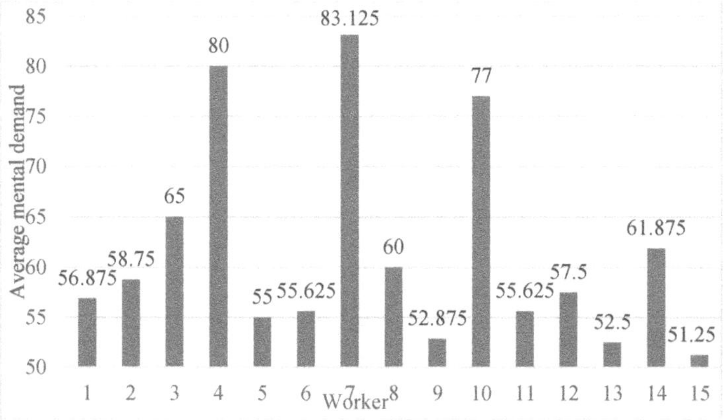

Figure 4.16 Average mental demand of the fifteen workers in the study

Mental demand in NASA-TLX is considered and analyzed. Mental demand is all about the mental and perceptual activity required to assemble the pens as per the scenarios. This mental demand also identifies whether the task was easy or demanding, simple or complex. The average mental demand of the fifteen workers in the study is illustrated in Figure 4.16. The mental demand for the two workers was very high, thus making it difficult for them to perform in all the scenarios. The remaining thirteen workers had almost similar mental demands except for worker 10. This variation in mental demand necessitates worker allocation to increase the assembly line's efficiency.

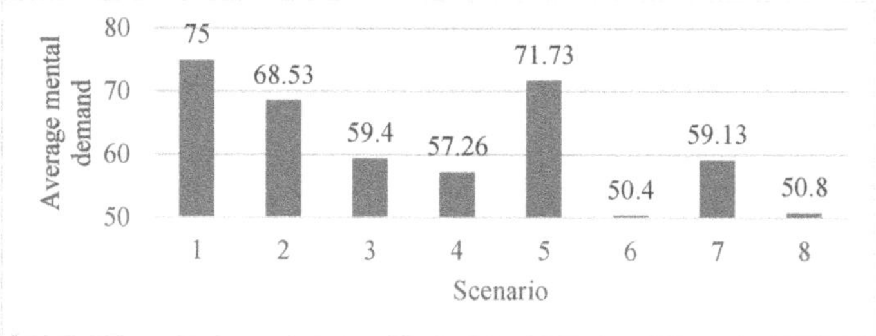

Figure 4.17 Average mental demand of the different scenarios in the study

The average mental demand of the fifteen workers of eight different scenarios is depicted in Figure 4.17. The first four scenarios are almost similar because the material presentation is similar and simple with the least mental demand. From scenario 5 the material presentation is getting complicated and the chance of assembling a pen without error is reduced. Part numbered box with instructions was the scenario in which a higher number of pens were assembled, so again the mental demand declines from scenario 5 to scenario 6. Scenario 8 has the lowest mental demand as the workers already learned about the parts and the process to assemble. Thus, a subjective measurement contributes to the experiment that the need for selection of different combinations according to the performance of each worker is to be considered.

4.8 Cognitive penetration and aesthetics

Burnston [41] claimed that the Openness / Intellect domain of personality and aesthetic engagement are related. Neither of these, though, are simplistic constructs and although the relationship exists, there is still a lack of process-based evidence justifying the relationship. By analyzing the effect of the facets of Openness and Intelligence on many different aesthetic emotions, this study sought to explain the relationship. In response to the number of a certain type of pen assembled, this study looked at the variations between and within-person in terms of emotions of interest and selection. Figure 4.18 illustrates the type of pens and the number of pens assembled. Subjects assembled Type A pens the most and Type F pens the least. This variation is due to the fact that type A pens are easy to distinguish; their color, reflection and the similarity outside made the workers consider that type of pens to be assembled first. Many of them successfully assembled, but the inner part of the pen such as the refill and the spring, were not similar. The effect of cognitive factors on sensory aspects of perception is known as cognitive penetration. Cognitive penetration typically occurs according to the evidence from neuroscience and philosophy, but to identify it in an experimental study is difficult. So, for the Type A pen for all the workers the cognitive penetration happened and that is the reason why it is not having the highest number of pens assembled without errors. The highest number of pens assembled without error is for type B. Type B pen is also aesthetically pleasing, but their inner parts are higher in number and similar to the outside parts. Cognitive penetration can be seen in all the types of pens assembled while comparing the number of pens assembled. The size and color of the spring and the refill are the toughest parts to identify and assemble. When the worker started from the first scenario to the last, they identified the pen parts based on instructions and photographs which might help them improve the number of pens assembled.

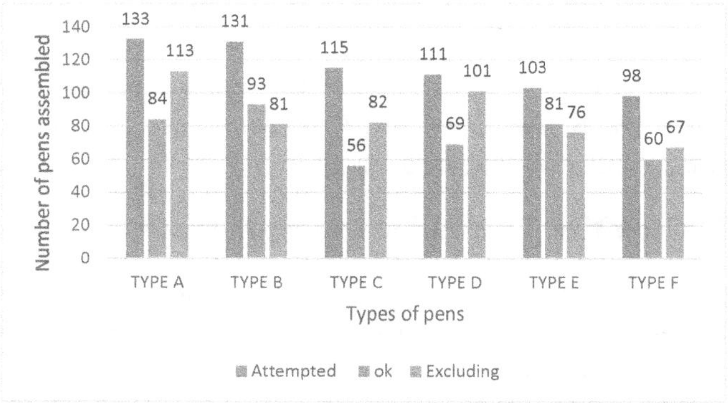

Figure 4.18 Pen types and total number of pens assembled

4.9 Development of predictive model – Logistic regression

With the results of the experiment, the worker allocation is done. A logistic regression model has been developed using the data obtained. This proposed model classifies the workers in terms of whether they will assemble the pen without error or not. The model is defined as given the time constraints for assembling a pen, predict if the subject will assemble the pen without error (0) or not (1).

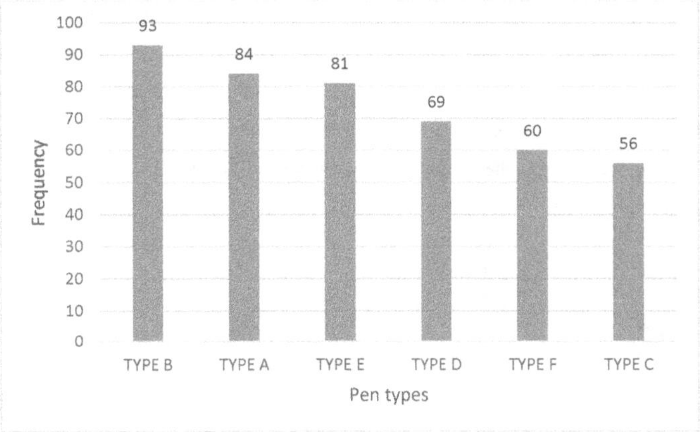

Figure 4.19 Frequency of pens assembled without error vs pen type

From figure 4.19 it is evident that if a worker is given Type B pen, then there is a higher chance that the worker assembles it without error, whereas for Type C there are chances of getting assembled but cannot assure the quality of the assembled pen.

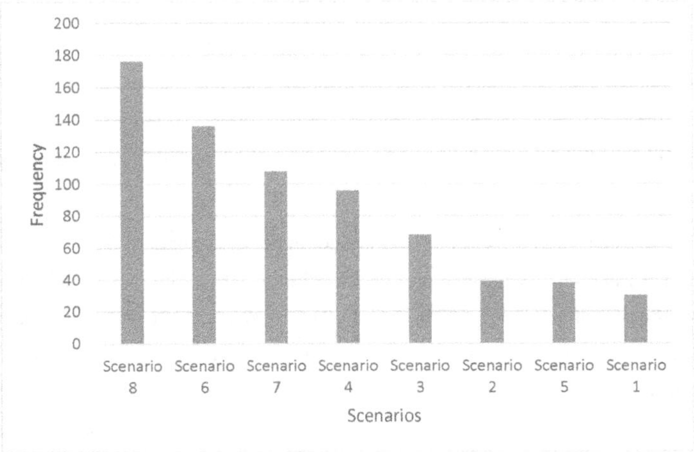

Figure 4.20 Frequency of pens assembled without error vs Scenario

Figure 4.20 depicts that in Scenario 8, there are higher chances that the workers assemble without error, and in Scenario 1, the chances of success are very low. This suggests that some workers are good in Scenario 1 but others are not. As a whole model, the chances of assembling the pen successfully without error from Scenario 8 to Scenario 7 is high. According to the workers, the frequency is illustrated in Figure 4.21.

Figure 4.21 Workers and the frequency of pens assembled without error

In this experiment, worker 6 is the best performer and worker 14, the slow performer. According to these results the worker allocation can be done. It is observed that the workers who had assembled Type B pens first and tried to assemble more pens in scenarios 8 and 6 (part numbered box with instructions and photograph) gave the best results.

The logistic regression model developed has an accuracy of 0.82, which can be seen in Table 4.6. The model is more specific than sensitive. The overall model could be improved with more data and by trying other classifiers. Moreover, in the statistical analysis of binary classification, F1 score is a measure of a test's accuracy, which is 0.82 in this case. It is calculated from the precision and recall of the test, where the precision is the number of correctly identified positive results divided by the number of all positive results, including those not identified correctly, and the recall is the number of correctly identified positive results divided by the number of all samples that should have been identified as positive.

Table 4.6 Performance metrics of logistic regression model

Recall score: 0.9158

Precision score: 0.7480

F1 score: 0.8235

	Precision	Recall	f1 -score	Support
0	0.91	0.74	0.81	125
1	0.75	0.92	0.82	107
Accuracy			0.82	232
Macro avg	0.83	0.83	0.82	232
Weighted avg	0.84	0.82	0.82	232

The summary of the results suggests that:

- The -Kruskal Wallis test suggests that part numbered box is the best way to present materials to workers and photograph is the best way to present information. Even though the statistical analysis verifies this, but in an ergonomic perspective this won't be acceptable. Each worker has his own acquired skills and abilities and inbuilt skills, so based on this any of the scenarios can be good for them depending on the cognitive aspects. It is preferable to analyze all the combinations of information and material presentation individually before arriving at conclusions. So, an ergonomic workplace must check these cognitive aspects of the workers which in turn improves productivity and adds value to the unit.
- The human performance measurements are experimental study data in which we use the number of errors and number of attempts. This is actually a detailed study of the different combinations of the cognitive aspects considered. The number of pens assembled in each scenario is important as when it comes to some workstations, it is important that the initial productivity contributes to the demand as the assembly process is continuous. For example, in this study

workstation 1 is complicated and once the activities in this workstation is completed, can be moved to the next workstations. In order to get the best worker for this job a detailed performance measurement is required. Another benefit of human performance measurement gives the workers with poor performance and they can be allocated to workstation 5 which requires less expertise.

- Cognitive penetration and aesthetics are linked to theories in cognitive science and exploring them helps to make the worker allocation effective. Even the parts color, shape, size etc. has an influence on each individual. When it comes to actual engineering drawings of a part the worker will face difficulty in assembling it, but when the characteristics of the part are identified and can relate it to a final product the worker will succeed in his job. Based on the experimental study we must consider the person with color blindness while designing the material and information presentation
- Development of predictive model helps the pen recycling unit to identify the right worker when a newly recruited worker joins and thereby identifying the workstation in which he/she fits.

4.10 Discussion on worker allocation

Hashemi-Petroodi et al. [7] presented an overview of workforce reconfiguration studies and the various strategies adopted for different manufacturing conditions as a part of workforce planning. Therefore, inspired by this research, a methodology for worker allocation in a pen recycling unit, an initiative of a non-profit organization in the southern part of India, is proposed in the present work. There is an analogy between scenarios in the experimental study and workstations in the pen-recycling unit with their tasks, abilities and skills required are all similar.

The pens collected for reuse will be in a heap mixed with different varieties in the initial stage. In the initial stage, it is necessary to disassemble them to remove defective parts and restore them with a new refill.

The tasks to be done in each workstation of the pen-recycling unit is listed below.

Workstation 1: Disassemble the pens collected and use only the good quality parts.

Workstation 2: Group the similar parts and arrange them based on part numbers or similarity in appearance.

Workstation 3: Assemble the pen with a new refill with instructions or photographs.

Workstation 4: Assemble the pen with a new refill with part numbers.

Workstation 5: Arrange and pack the assembled unit.

There is an analogy between scenarios 1,3, and workstation 1 and scenarios 2,4 and workstation 2. In the first four scenarios, the worker does not even have an idea about the pens other than the material and information presentation given; the same situation happens when you are introduced to a heap of pens and choose the non-defective parts. There is an analogy between scenarios 5,7, and workstation 4 and scenarios 6,8 and workstation 3. Here, the scenarios are provided with instructions and photographs, but part numbers are the best option for some workers to assemble the pen.

The five workstations of the pen-recycling unit are depicted in figure 4.22. A comparative analysis of information and material presentation has been a primary objective. Based on the analysis of results, about fifteen workers need to be allotted to these work stations. The workers are allotted based on their performances in experimental study. This helps to identify the worker skills and assign them to the right work stations.

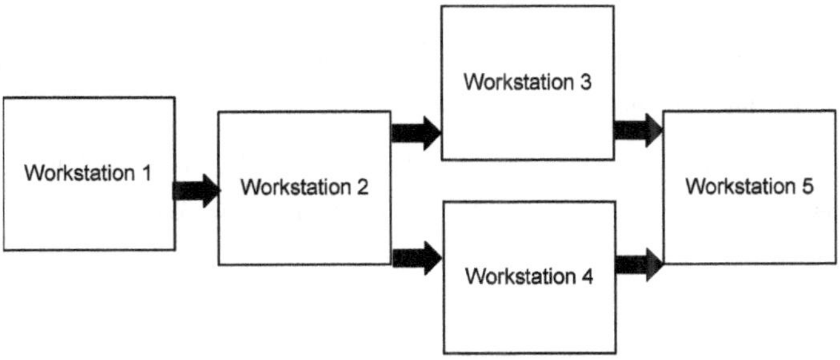

Figure 4.22 Workstations to which allocation is to be done based on experimental results

The allocation of workers is based on the number of pens assembled by all the workers in each scenario. The scenarios in which maximum and minimum number of pens assembled by the workers is depicted in Table 4.22.

Scenario 1 and 3 in the experimental setup matches with work station 1. Therefore, for the allocation of workers to the first workstation, the following two conditions are to be met.

Condition 1: The worker considered must have maximum number of pens assembled in scenario 1 and 3

Condition 2: The worker considered must not have minimum number of pens assembled in either scenario 1 and scenario 3

If the first condition is not satisfied and the second condition is satisfied, then those workers can be considered as an initial allotment. Hence, workers 3,4,5,7 and 14 are considered.

Scenario 2 and 4 in the experimental setup matches with work station 2. For the allocation of workers to the second workstation, the following two conditions have to be met.

Condition 1: The worker considered must have maximum number of pens assembled in scenario 2 and 4

Condition 2: The worker considered must not have minimum number of pens assembled in either scenario 2 and scenario 4

If the first condition is not satisfied and the second condition is satisfied, then those workers can be eliminated as an initial allotment. Hence, workers 1 and 6 can be considered.

Scenario 5 and 7 in the experimental setup matches with work station 4. For the allocation of workers to the fourth workstation, the following two conditions are to be met.

Condition 1: The worker considered must have maximum number of pens assembled in the scenario 5 and 7

Condition 2: The worker considered must not have minimum number of pens assembled in either scenario 5 and scenario 7

If the first condition is not satisfied and the second condition is satisfied, then those workers must be eliminated. Hence, workers 3,5,6,7,8, and 15 can be considered

Scenario 6 and 8 in the experimental setup matches with work station 3. Hence, for the allocation of workers to the third workstation, the following two conditions must be met.

Condition 1: The worker considered must have maximum number of pens assembled in the scenario 6 and 8

It is observed that most of the workers satisfy this condition and have maximum number of pens assembled either in scenario 6 or 8. Therefore, the maximum number of pens assembled in either scenario 6 or 8 are considered for selection of workers to workstation 3. Hence, workers 10,11, and 12 are allocated as their respective output being 14, 14 and 19 as per Table 3.

For the fifth station, the workers who are less skilled and not included in any of the four work stations can be considered. Therefore, the workers allocated will be 2, 9 and 13 to the final workstation. The data of these worker allocation is tabulated in Table 4.7.

Table 4.7 Stations and expected workers in each station

Work Stations	Recommended Workers
1	3,4,5,7,14
2	1,6
3	12,11,10
4	3,5,6,7,8,15
5	2,9,13

Each workstation requires three workers each out of the total 15 workers. First, we consider the workers who are suitable only for a single workstation. For station 3 and 5 exact number of workers is already allotted as per Table 8. Workers 12,11 and 10 can be assigned to station 3 and workers 2, 9 and 13 can be allotted to station 5. Station 2 can be assigned only to worker 1 and 6, though three workers are required. For station 4, workers 3, 5 ,6 ,7 ,8 ,15 can be allocated. Since worker 6 is already allocated to station 4, the only options left is 3, 5, 7, 8 and 15. Out of the five workers for station 4, workers 8 and 15 are exclusive. Workers 3, 5 and 7 are common to both station 1 and 4 so thereby they can be reallocated according to the requirements.

In stations 4 and 1, workers 3, 5 and 7 can be allotted as they are suitable for both the workstations. But in comparison with the number of pens assembled in scenario 5 and 7 (at Station 4) with that of scenario 1 and 3 (at Station 1), as per Figure 4.11 and 4.12, worker 7 can be confirmed. Worker 7 has assembled a greater number of pens and can be allocated to station 4. For station 1, already workers 4 and 14 are allocated. The workers 3 and 5 are yet to be allocated. Either worker 3 or worker 5 can be allocated to station 1 and the remaining one worker must be trained to accommodate in station 2. As worker 3 has the greater number of pens assembled compared to worker 5 in scenario 1 and 3 (see Figure 4.11), worker 3 is preferred. Further worker 5 is remaining and needs to be trained to accommodate him/her in station 2 to satisfy the workforce. Hence training must be given only to a single worker according to the cognitive aspects. This reduces the cost of training the workers and allows for effective allocation of workers to their respective workstations with which they can achieve maximum output without errors. The final allocation of the workers is listed in Table 4.8.

Table 4.8 Final allotted workers to each station

Work Stations	Final Allotment of Workers
1	3,4,14
2	1,6,5
3	12,11,10
4	7,8,15
5	2,9,13

The most critical task which can be handled by efficient workers increases the rate of recycling. The assembly unit will be operational twice in two weeks. Thus, a final allotment of the fifteen workers are effectively done based on the results from the experiment. Both instructions and photographs can be provided in station 3 and 4 for the convenience of the workers. This method of allocation can be designed based on the assembly lines, and it varies from worker to worker.

4.11 Case study: Pen recycling unit

The purpose of this case study is to enhance the performance of the workers in a pen recycling unit. A previous work with a pilot study was success in the laboratory environment. In order to verify the methodology adapted this is applied for pen recycling unit. The pen recycling unit as a social institution lacks funds but the amount of the pens to be recycled was abundant. The school and college students nearby contributed a lot of used pens to the recycling unit. There came a situation that these pens were not able to be recycled properly and also space issues which lead to closing of the unit. There is a huge demand among the students as they get these pens for a subsidized rate. A visit to the unit was done and identified the pen recycling unit. CLAM factors were used to adapt a scientific methodology to the study. The main issue with the recycling unit is the variety of products with different part sizes and colors which confuse the workers and leads to cognitive load. The possible solutions to this issue was listed as follows:

- Recruit highly skilled workers
- Increase the number of workers to 30
- Layout improvement, rack setup and extension of the unit
- Guidelines preparation for information and material preparation
- Worker allocation based on performance.

The first three solutions were not at all economical and we couldn't consider that because they can't allocate adequate fund. So, we concentrated on information presentation and material presentation improvement in the pen recycling unit. Then the pilot study which we conducted in our lab was conducted with the same setup. Then the experiment design was correlated with the workstations of the pen recycling unit and analysis was done. The person with the right skill was selected and allocated to the workstations. Then again, the workers work in the allocated workstations and their performance is monitored using daily hourly monitoring. An increase in the number of pens assembled is achieved when the proposed methodology is utilized. To present the proposed approach systematically a flowchart is given in Figure 4.23.

A further increase in the productivity can be achieved by improving cognitive abilities by providing trainings and by providing short breaks to reduce the cognitive load (was effective during the actual study). This study can also be considered as a verification of the pilot study designed. The hypotheses considered gave similar results in the pilot study conducted with the workers and the pilot study conducted in the lab environment. The allocation of workers with the methodology adopted in this study improved the productivity. This data was cross-checked with the company officials and the before and after chart is added and depicted in Figure 4.24.

From the data collected for eight hours, it is observed that at the end only 10 pens are assembled correctly before applying the methodology but now it has increased to 40. A four times increase can be seen. Here the skill is so important because the first workstation is critical as they have to be highly skilled to disassemble the pens collected and use only the good quality parts. So, the workers in first workstation must be able to distinguish among the parts and they are also responsible for maintaining the quality and reducing errors. The workstation follows a continuous assembling process as all the workstations must have skilled workers to achieve a good count of number of pens assembled.

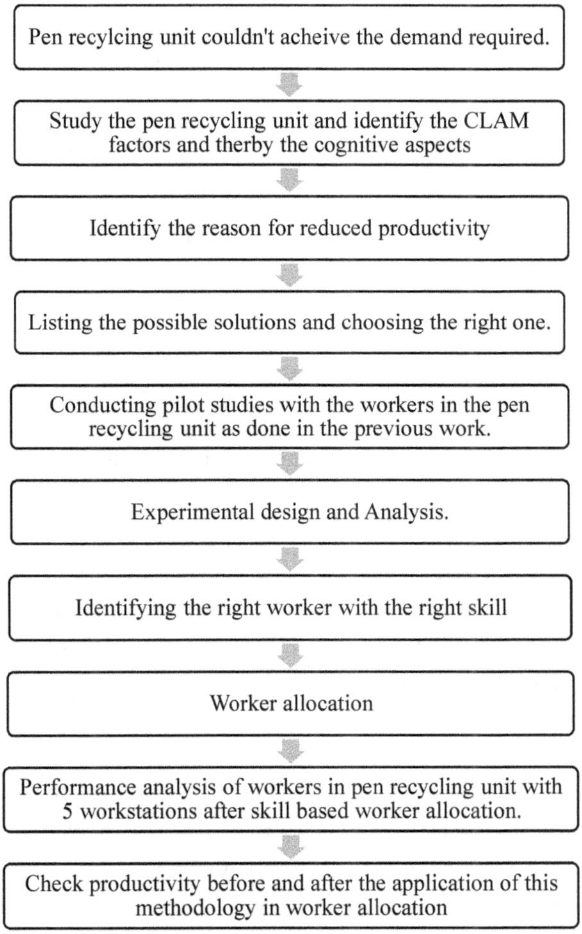

Figure 4.23 Flowchart of the proposed methodology in pen-recycling unit

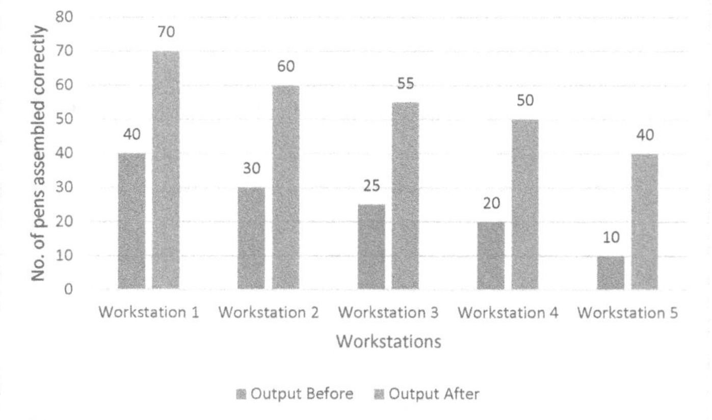

Figure 4.24 Variation of results before and after applying the proposed methodology

4.12 Summary of Chapter 4

The study here presented concerns the theme of cognitive aspects to be considered in assembly line, still little explored in literature in terms of worker performance, skills and their importance in worker allocation. The study depicts how significantly worker performance in manual assembly systems differs from what is typically accounted for by workstation designers, and how such differences might be linked to cognitive aspects. The significant differences found in worker's performance are due to the varying cognitive capabilities of the workers. The influence of cognitive aspects namely, information presentation and material presentation on individual human performances are identified. Problems that arise from poor workstation design of manual assembly processes may entail cognitive workload, leading to less effective working and, therefore, a decrease in production efficiency. To address this issue, a real-time case study in a pen-recycling unit with a methodology is proposed. The methodology is aimed at identifying the right worker for the right workstation through a pilot study. This approach is simple and economical, and can be applied to all manual assembly workstations with a wide variety of products. The methodology reduces the complexities in identifying the objective function, formulating an objective function and the constraints when the optimization techniques are applied for real-time case studies for worker allocation (based on skill). In essence, this work can be summarized as follows:

- Cognitive aspects affecting human performance in manual assembly have to be considered to improve productivity and reduce errors.
- The cognitive aspects can be selected from CLAM factors according to the workstation to be studied.
- All possible combinations must be included in order to obtain the best results (For example, this research took all possible combinations of material and information presentation.
- Worker allocation is an important factor which decides the output in a workstation.
- Skill-based worker allocation is preferred.
- By following the methodology proposed, better results can be obtained.

The analysis of results says that the performance of each worker improved when they were allocated to the workstations based on the cognitive aspects. The results revealed that the methodology could be successfully adapted, but the pilot study needs to be varied according to the requirements of the production unit. The proposed methodology is therefore designed for industrial sectors in which assembling operations involve a wide variety of products and subparts which require assembling the products without errors. In these companies, a skilled workforce is a primary resource and, like any other, is a limited resource. A skilled workforce increases costs, but study suggests identifying workers who are already skilled and those who can be trained to become skilled. In future works, the method could be enhanced by implementing this methodology in workstations which utilize some tools, robots etc. The cognitive abilities to be considered in the pen assembly task can be determined using the virtual cognitive task. The next chapter discusses an experimental study with eight cognitive abilities using virtual cognitive task.

Chapter 5

Identifying the cognitive abilities using virtual cognitive task

5.1 Introduction

Cognitive research, including psychology, philosophy, linguistics, neuroscience, computer science, and anthropology, may be considered a union of various fields. Cognitive scientists strive to understand the essence of our thought, what is universal and variable about it, and how the variety of situations we encounter reacts to and is influenced by human cognitive mechanisms [59]. Cognition is a term that refers to the mental processes involved in learning and understanding. Many aspects of intellectual functions and processes, such as attention, memory, computation, decision making and learning are also covered in cognition. According to Barber et al. [103], a controversy persists regarding cognitive skills and their connection to human variations, particularly in the context of cognitive abilities informed by future time perspective. Palanisamy et al. [104] explained certain tests and assessments which are used for determining cognitive status of individuals. Both these methods are subjective measures based on a questionnaire, and the accuracy of the data collected needs to be rechecked. Out of this cognitive research topics an interesting concept is the cognitive abilities based on [6]. The cognitive abilities and the skills involved helps to design certain cognitive tests. The cognitive tests are done through brain blog research websites which integrate brain health into training, leadership programs etc. There is growing consensus among cognitive neuroscientists that meaningful progress in understanding human cognitive skills depends on enhanced data acquisition as well as coordinated data synthesis and integration across multiple studies and research laboratories. Efforts of Yarkoni et al. [60] for an increased focus on cognitive skills and their effective way of identification of tasks and experimentation using cognitive atlas helped the current study for defining the experimental design.

The mental processes that help us to receive, select, store, transform, develop, and recover information that we have obtained from external stimuli are brain cognitive functions. Cognitive abilities are brain-based abilities required from the easiest to the most complex to carry out any task. They are connected to the processes of how we read, recall, solve and pay attention to problems, etc. This research investigates how various cognitive abilities influence the cognitive skills of the workers by giving them tasks based on the cognitive test and giving scores for all eight cognitive abilities identified. Cognitive abilities are to be considered to identify the variation of easiness to complexity while carrying out any task.

The cognitive abilities considered for skill assessment in the present study are as follows:

- Visual perception
- Attention
- Memory
- Motor skills
- Language
- Visual and Spatial processing
- Executive functions and
- Learning.

The total score of the above-listed eight cognitive abilities for all thirty subjects are noted. The weights of the cognitive abilities are varied, and thus the influence of these abilities on the total score of the subject is analysed using statistical analysis A virtual cognitive test typically refers to an activity or experiment designed to assess various cognitive abilities is named as virtual cognitive task in this research. These tasks are often used in psychological research or cognitive assessments, and with advancements in technology, many of these tasks have been moved to virtual platforms. The cognitive abilities and skills used in this study are explained in table 5.1 below:

Table 5.1 Cognitive abilities and skills involved

Cognitive Abilities	Skills involved
Perception	Recognition and interpretation of sensory stimuli.
Attention	Ability to sustain concentration on a particular object, action, or thought, and ability to manage competing demands in our environment.
Memory	Short-term/ working memory (limited storage), and Long-term memory (unlimited storage).
Motor skills	Ability to mobilize our muscles and bodies, and ability to manipulate objects.
Language	Skills allowing us to translate sounds into words and generate verbal output.
Visual and Spacial processing	Ability to process incoming visual stimuli, to understand spatial relationship between objects, and to visualize images and scenarios.
Executive functions	Abilities that enable goal-oriented behaviour, such as the ability to plan, and execute a goal.
Learning	Ability to learn and reduce errors made earlier. A learning processes.

The virtual cognitive task is an experimental study done with thirty subjects and this methodology can also be applied to the workers in the industrial environment to identify their cognitive abilities.

5.1.1 Advantages of Virtual cognitive tasks

There are various advantages of these virtual tasks which makes it a perfect cognitive test for studies in the manual assembly. These studies were performed during the lockdown which helped to continue our research smoothly in online mode. Some of the advantages are listed below:

1. Accessibility: Virtual tasks can be accessed remotely, making them ideal for large-scale studies or assessments conducted across different locations.
2. Controlled Environment: Virtual tasks allow for precise control over the testing environment. For example, external distractions can be minimised, and stimuli can be presented at controlled intervals.
3. Data Precision: The digital nature of virtual tasks provides highly accurate performance data. For example, reaction times, error rates, and response patterns can be tracked with great precision.
4. Engagement: Virtual tasks can be designed to be more engaging and interactive than traditional paper-and-pencil tasks, which can improve participant motivation and compliance.
5. Scalability: Virtual platforms can easily accommodate large numbers of participants, making it feasible to conduct studies on diverse and larger populations.

5.1.2 Applications of Virtual Cognitive Tasks

The virtual cognitive tasks help in experimental studies of manual assembly tasks as they can provide the performance of workers based on their cognitive abilities. The applications of virtual cognitive tasks are as follows:

1. Workstation Assessments: Virtual cognitive tasks can be used in workstations of manual assembly work environment where assembling of small parts is required.

   The workers who are not eligible to perform a certain task in a workstation can be identified and a re-allocation of workforce can be done.
2. Education and Training: Virtual cognitive tasks can be used in industrial environment to help workers develop cognitive skills. The workers who are not

fit in any of the workstations can be given such trainings so that the chances of errors made will be reduced.

5.2 Participants

Thirty subjects aged between 20 and 30 years of age volunteered for the study, with a mean age of 21.93 years and a standard deviation of 3.23 years. Subjects were given all the instructions prior to the task. All the subjects had done these tasks in the comfort of their home in online mode. The experiment was conducted in a friendly atmosphere to avoid the stress and to obtain maximum output.

5.3 Experiment Procedure & Apparatus

There are eight cognitive abilities to be tested for all the subjects. The cognitive tests are done through brain blog research websites which integrate brain health into training, leadership programs etc. A cognitive test where the scores of all eight cognitive abilities where defined was selected. Each subject performed different tasks/ responded to the questions asked respective to the eight scenarios and equal time were given to all.

The study was conducted with two laptops where one will be on the instructor's side and the other will be on the subject side. The subject and the instructor are in their respective homes and the internet speed test was conducted before the experiment. A fixed amount of time was given to all the subjects and one to one interaction was done through an online video calling platform. Eight cognitive abilities of the subjects are conducted with eight different sections. All the instructions for the tasks were illustrated using the screen sharing method as per Fig. 5.1 Instructor also gives oral instructions and the scores are noted manually.

Figure 5.1 Virtual view from the instructor's laptop during the task

5.4 Experimental design

A person's cognitive abilities impact job performance. Therefore, cognitive ability plays a crucial role in an employee's job suitability and success. Subjects who score well on cognitive ability tests are more likely to complete the training successfully, learn and digest new information on the job, and adapt more quickly in fast-changing work environments [61]. Eight cognitive abilities are considered in the study. For each cognitive ability a brain teaser or questions based on the section are displayed on the sharing screen of laptop. The subject responds and the scores are recorded. Mostly brain teasers are given in each section. The individual scores and their weightages are shown in Table 5.2. The scores added up to a total of 73. This score is best suited and is scientifically proven which is provided by brain blog research websites (sharpbrains).

Table 5.2 Cognitive abilities and their scores

Cognitive Ability	Score	Cognitive Ability	Score
Visual Perception (VP)	8	Language	10
Attention	7	Visual & Spatial processing (VSP)	9
Memory	11	Executive functions (EF)	16
Motor skills	2	Learning	10

5.5 Results and Discussion

All the thirty subjects performed the tasks and their scores out of 73 was noted. The performance of the subjects based on their cognitive ability scores are listed in table 5.3 given below. Subject 30 has the highest score and subject 5 has the lowest. The performance rating is done based on the individual scores obtained for each cognitive ability.

Table 5.3 Subjects and total score of 8 cognitive abilities

Subject	Cognitive ability score	Subject	Cognitive ability score
1	45	16	34
2	44	17	48
3	41	18	43
4	39	19	39
5	32	20	44
6	48	21	43
7	52	22	47
8	50	23	36
9	44	24	47
10	42	25	41
11	46	26	39
12	34	27	38
13	46	28	43
14	41	29	38
15	50	30	56

The scores are further divided based on their performance into class intervals as per the Table 5.4. Six out of thirty subjects could achieve the best scores in the study. Six of the subject's performance was poor and eighteen of them performed good in the experiment.

Table 5.4 Class intervals and frequency of the cognitive ability scores

Class Interval	Frequency
Poor (30 -38)	6
Good (39-47)	18
Very Good (48- 56)	6

5.5.1 Wilcoxon test

The fixed score of 73 from brain blog research websites needs to be statistically tested and need to identify whether the rankings will be the same when the cognitive ability scores are given different weights. The differences in the scores of the two matched groups similar to the *t*-test for two related samples are done using Wilcoxon Signed Ranks Test. Equal weightages score is out of 80 and unequal weightages score is out of 73.

- **Null hypothesis, H0:** Rankings are the same
- **Alternative hypothesis, H1:** Rankings are different

[DataSet0]

Descriptive Statistics

	N	Mean	Std. Deviation	Minimum	Maximum
EQUALWEIGHTAGES	30	53.7500	6.95583	40.00	70.00
UNEQUALWEIGHTAGES	30	54.6119	8.54899	35.62	75.34

Wilcoxon Signed Ranks Test

Ranks

		N	Mean Rank	Sum of Ranks
UNEQUALWEIGHTAGES - EQUALWEIGHTAGES	Negative Ranks	11[a]	14.36	158.00
	Positive Ranks	19[b]	16.16	307.00
	Ties	0[c]		
	Total	30		

a. UNEQUALWEIGHTAGES < EQUALWEIGHTAGES
b. UNEQUALWEIGHTAGES > EQUALWEIGHTAGES
c. UNEQUALWEIGHTAGES = EQUALWEIGHTAGES

Test Statistics[a]

	UNEQUALWEIGHTAGES - EQUALWEIGHTAGES
Z	-1.533[b]
Asymp. Sig. (2-tailed)	.125

a. Wilcoxon Signed Ranks Test
b. Based on negative ranks.

Figure 5.2 Screenshot of Wilcoxon output from SPSS

As per figure 5.2, $P > 0.05$ so thereby failed to reject the null hypothesis.

This suggests that the rankings will be the same when the cognitive ability scores are given different weights.

5.6 Summary of Chapter 5

Virtual cognitive tasks are powerful tools for assessing, understanding, and improving cognitive abilities. The study suggests a cognitive test in a virtual environment and the total score of the cognitive abilities for thirty subjects are listed. The individual performs a few tasks in a specific cognitive ability but failed to perform in other, makes a difference in the score. Subject 30 (56 out of 73) has the highest score and subject 5 (32 out of 73) has the lowest. Feedback of the score for each cognitive ability can be made available and can train the subjects Identifying the cognitive abilities of subjects helps to fit the workers for the right task in the right workstation. The next chapter discusses the allocation of workers using personalised models obtained from the total cognitive ability scores and the pen assembly task.

Chapter 6

Worker allocation using personalised models from cognitive abilities

6.1 Introduction

Modern manufacturing systems are mostly automated with robots and use the shop floor management cycle with pneumatic and hydraulic circuits to make the task easier in an assembly line. Even in the industry 4.0 era, there are some workstations in advanced manufacturing units where small parts are being assembled manually. This research is on such workstations, and reports on the small error in such workstations which may lead to a lot of reworks. Any business enterprise aims to expand its manufacturing system, but with the large variability of manufactured products, the system becomes more complicated and not economical as per Dolgui and Proth [105]. Hence in such systems, where operators have to work with more variety of components, the concept of cognitive ability is to be assessed. This study reports a novel methodology of allocating workers based on their performance and cognitive abilities. This study investigates how various cognitive abilities influence the pen assembly experiment of the subjects when compared to their performance in the workstation. The cognitive abilities of workers, their ability to perform multiple tasks in various locations with various functionalities (such as Quality controller and Assembler) are identified. The cognitive aspect of the study consists of visuospatial information processing and storage. The pen assembly task done by different subjects in the lab environment can be considered as an analogy of the workers in a manual assembly in the industrial environment. The virtual cognitive task is an experimental study done with thirty subjects to identify their cognitive abilities. The combination of pen assembly task and virtual cognitive task provides the cognitive abilities required for a subject to complete the pen assembly task without errors. Similarly in an industrial environment this methodology helps to identify the cognitive abilities required for the worker which helps in worker allocation

Cognitive Load Assessment for Manufacturing method (CLAM) method helps in identifying and assessing the occurrence of cognitive load in manufacturing personnel. Educating and assisting manufacturing companies to reduce cognitive load in the personnel at the shop floor early can also be achieved. The CLAM factors and their descriptions are listed in table 3.1 of chapter 3.

NASA Task Load Index (NASA-TLX) is used in this study to measure cognitive load. The present study also tried to prove that the mental demand component of the NASA-TLX closely aligns with the cognitive load, as mentioned by the authors. The self-reported cognitive load is related to physiological data, which enables it to measure cognitive load in real-time and without interrupting the subject through directly measurable variables [53]. One such method is used in the study known as Instantaneous self-assessment (ISA). The ISA is a self-rating method of the current mental workload during a task on a scale of 1 (low) to 5 (high). It is normally queried every 2-5 minutes during the activity.

According to Jercic et al. [51] an underlying hypothesis states that heart rate is directly linked to physical activity but is also a measure of nervous system activity, revealing the indirect link of heart rate with cognitive load. Solhjoo et al. [52] demonstrated that stress or arousal will cause an increase in blood pressure and a decrease in heart rate variability, thereby identifying a link with cognitive load. Changes in various bodily processes and states have also been reported with changes in mental workload. One major advantage of psychophysiological measures is the continuous availability of bodily data, allowing load to be measured at a high rate and with a high degree of sensitivity, even in situations in which overt behaviour is relatively rare and explains about heart rate as a Psychophysiological measurement as per Paas [13]. However, Brunken et al. [49] explains psychophysiological measures are also very sensitive to physical effort and will reflect specific mental load variations only for activities involving little or no physical effort. Several measures can be used to estimate mental workload: cardiac activity, electrooculogram, respiration or event related potentials.

6.2 Methodology

The experimental part is the pen assembly task and virtual cognitive task, which is already discussed in chapter 4 and 5 respectively. This study is focused on performance of the individuals in the pen assembly, by identifying the required cognitive abilities for assembling the pen and thereby worker allocation is done.

The pen assembly task in chapter 4 has to be recalled here. For eight scenarios in pen assembly task is converged to 4 tasks. The CLAM factors are applied. The hypothesis study is done and results are obtained. As a subjective workload assessment NASA-TLX and ISA is used and for psychophysiological measurement, heart rate is considered. Statistical analysis on the data is done and results are listed. The worker allocation flow chart is shown in figure 6.1

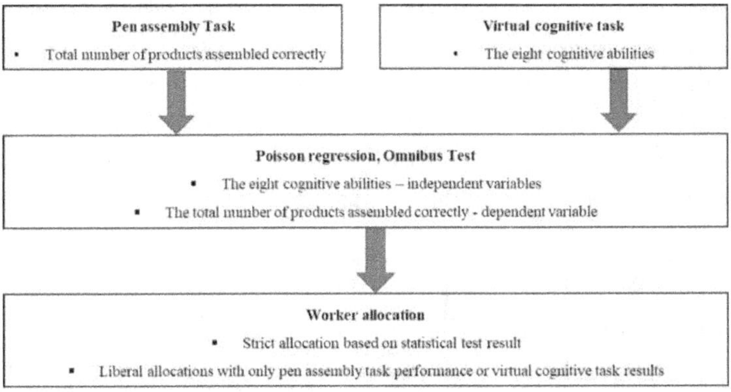

Figure 6.1 Flowchart of the methodology

6.3 Participants

In total, 30 subjects aged between 20 and 30 years volunteered for the study, with a mean age of 21.93 years and a standard deviation of 3.23 years. Chapter 1 only has 15 subjects are but pen assembly task was further done for 30 subjects. The same thirty subjects who participated in pen assembly task has done the virtual cognitive tests also.

6.4 Experimental design and tasks

The different scenarios explained in table 4.1 of chapter 4 needs to be considered here. Those scenarios are grouped based on the complexity of the tasks. Task 1 will be combined study of scenario 1 and 3 which means with both instructions and photograph the subject has to assemble the material presentation of Intermixed parts. Similarly for task 2 it will be combination of scenario 2 and 4, task 3 will be combination of scenario 5 and 7, task 4 it will be combination of scenario 6 and 8.

To identify the cognitive load across different conditions, Sweller [106] explained two processes from the working memory, namely storing (remembering) information and processing information. Earlier, Kyllonen and Christal [107] had identified the good measures for working memory as simultaneous processing and storage, not involving learning, and requiring knowledge that all subjects are presumed to have. Considering the above facts from the mentioned authors, as well as the size, shape, and texture of the products at the actual workstation, tasks for this study are designed and conducted. The first method to vary cognitive load is by manipulating (visuospatial) information processing through the difficulty of the task. For that purpose, the difficulty of tasks is increased by mixing up the parts of the products to be assembled.

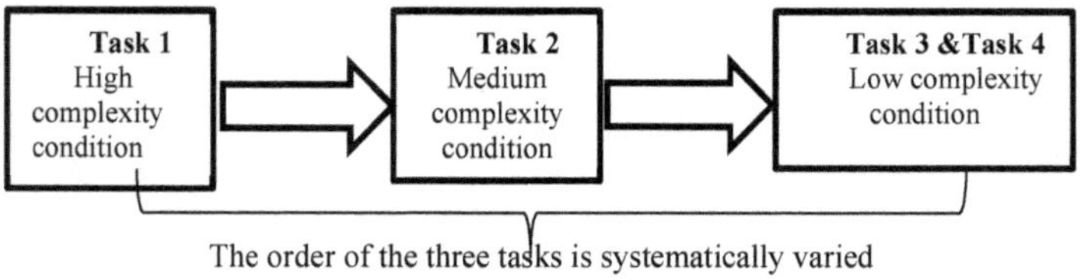

Figure 6.2 Flow chart of design and process of study

A second method to vary cognitive load is by manipulating information storage, and this is done by presenting information as stimuli in two different ways as instructions and images. The three different complexity conditions, high, medium, and low are designed bases on the small assembling workstation of the medium-sized enterprise. The tasks considered are characterized based on the complexity of conditions and are depicted in figure 6.2.

There are six products to be assembled, each with a different number of parts, as shown in Table 6.2. Two sets of products (1 to 6) are used for the study.

- In Task 1, all 62 parts (31 parts per set × 2 sets) of the 12 products are mixed, and the subject must assemble these products using two different information presentation methods (digital instructions and photographs). This is a high complexity condition as the variety of products varies with its texture and color.
- Task 2 involves a racking concept in which similar parts, such as springs, washers, and bushings for all six products, are grouped together. A total of 62 parts (31 parts per set × 2 sets) from 12 products are arranged in different racks, with two different information presentation methods (digital instructions and photographs).

Table 6.1 The product varieties and the list of the parts

PRODUCT 1	PRODUCT 2	PRODUCT 3	PRODUCT 4	PRODUCT 5	PRODUCT 6
PART 11	PART 21	PART 31	PART 41	PART 51	PART 61
PART 12	PART 22	PART 32	PART 42	PART 52	PART 62
PART 13	PART 23	PART 33	PART 43	PART 53	PART 63
PART 14	PART 24	PART 34	PART 44	PART 54	PART 64
PART 15	PART 25		PART 45		PART 65
	PART 26				PART 66
					PART 67

- Task 3 also follows a systematic rack concept, where parts are grouped based on a single product. Here the number of parts of a single product is increased to five. For example, in product 1 there will be five parts for part 1 with different colors and textures, altogether by assembling a total of 30 products will be obtained.

- Task 4 is also the rack concept wherein the traditional way of numbering the parts with their respective product families. The part labelling will be written on the racks as per table 3 (for example PRODUCT 2 will have PART 22) to distinguish between the parts and their respective product.

The main aim of the study is to cluster the subjects based on their skill set to be in the Quality control and the remaining who can be accommodated as assemblers and remaining subjects who require training. The total number of products assembled correctly in tasks 1 and task 2 determines whether the worker can be good at Quality control. The total number of products assembled correctly in tasks 3 and task 4 determines whether the worker can be good as an Assembler. Assembler and quality control is identified as different skill rating required (eg: 0 or 1) for the workers.

To give some technical concepts in the production units, this work has considered the Cognitive Load Assessment for Manufacturing (CLAM) method from Thorvald et al., [108] helps in identifying and assessing the occurrence of cognitive load in manufacturing personnel. Educating and assisting manufacturing companies to reduce the cognitive load in the personnel on the shop floor early can also be achieved. Out of 11 CLAM factors, four of them were considered.

- Variant flora (Increases from task 1 to task 4)
- Level of difficulty (Decreases from task 1 to task 4)
- Production awareness (Increases from task 1 to task 4)
- Quality of instructions (Increases from task 1 to task 4)

6.5 Hypotheses

The hypotheses used for this factorial experiment (HA, HB, HC, HD, HE, HF) are as follows:

- **Null hypothesis, H0A:** No significant difference between the performance when task 1 in high complexity condition and the performance when task 2 in medium complexity condition.

- **Alternative hypothesis, H1A:** Significant difference is observed between the performance when task 1 in high complexity condition and the performance when task 2 in medium complexity condition.
- **Null hypothesis, H0B:** No significant difference between the performance when task 3 in low complexity condition and the performance when task 4 in low complexity condition.
- **Alternative hypothesis, H2B:** Significant difference is observed between the performance when task 3 in low complexity condition and the performance when task 4 in low complexity condition.
- **Null hypothesis, H0C:** No significant difference between the performance when task 1 in high complexity condition and the performance when task 3 in low complexity condition.
- **Alternative hypothesis, H1C:** Significant difference is observed between the performance when task 1 in high complexity condition and the performance when task 3 in low complexity condition.
- **Null hypothesis, H0D:** No significant difference between the performance when task 1 in high complexity condition and the performance when task 4 in low complexity condition.
- **Alternative hypothesis, H1D**: Significant difference is observed between the performance when task 1 in high complexity condition and the performance when task 4 in low complexity condition.
- **Null hypothesis, H0E**: No significant difference between the performance when task 2 in medium complexity condition and the performance when task 3 in low complexity condition.
- **Alternative hypothesis, H1E**: Significant difference is observed between the performance when task 2 in medium complexity condition and the performance when task 3 in low complexity condition.
- **Null hypothesis, H0F**: No significant difference between the performance when task 2 in medium complexity condition and the performance when task 4 in low complexity condition.

- **Alternative hypothesis, H1F:** Significant difference is observed between the performance when task 2 in medium complexity condition and the performance when task 4 in low complexity condition.

6.6 Results

Every employer of a production unit is interested in the number of products which is produced without errors. Similarly, this study checks the products which are ready to be sent to the consumer. Out of the four tasks, task 4 is easier and designed scientifically, but comes under low complexity conditions. Task 3 also matches with task 2, which is a medium complexity condition. This refers to the importance of the two different tasks in low complexity conditions. Figure 6.3 represents the total number of products produced without errors in each task and figure 6.4 represents the errors made for the four tasks.

Figure 6.3 Count of products assembled correctly without errors in each task.

From the collected data the performance is identified using the monitored parameters, number of attempts, number of errors, and completion time.

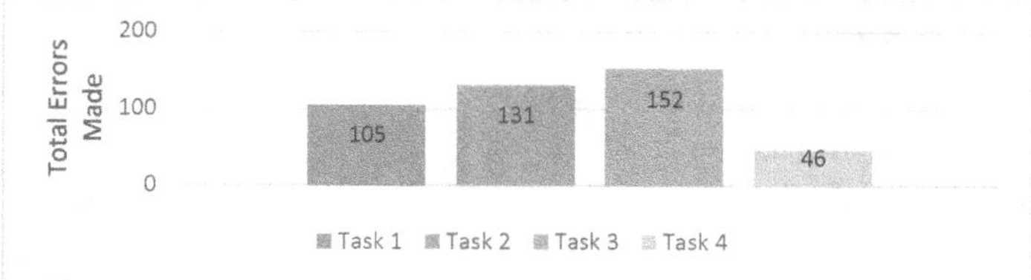

Figure 6.4 Total errors made in each task.

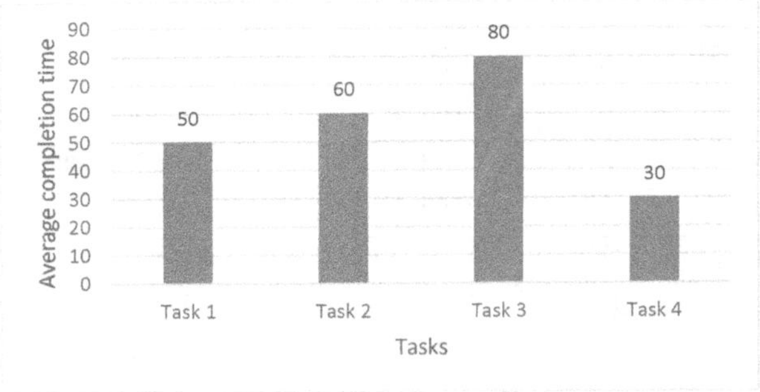

Figure 6.5 Average completion time in each task

The total number of products assembled correctly is highest and the number of errors made is lowest for task 4. Task 1 has the least number assembled but the number of errors made is lesser compared to task 2 and task 3 (variability, difficulty, learning effect, etc.). As depicted in figure 6.5, the average completion time for task 4 is lesser and is considered the best compared to task 3. Task 1 is having the highest level of difficulty, but since the tasks are given randomly there is a learning effect contributing to the high complexity condition. The number of attempts is found to be more in tasks 1, 2, and 3, but for task 4 the number of attempts is very less. All the performance measures suggest task 4 is better and subjects can perform well. In actual workstations, a person who is good in task 4 cannot work in quality control as his expertise is to follow a particular pattern to assemble according to the numbers. For people who are not good with numbers task 3 provides a performance measurement.

6.7 Hypothesis for product assembly task

From the initial analysis and the testing of the hypothesis, some of the main effects are significant in the analysis of the results. Examination of content presentation, using a comparative analysis in table 6.3, reveals more details about the performance between various tasks (based on p-value).

Table 6.2 Summary of the hypothesis test

Hypothesis	p-value	Decision
A	0.000	Reject the null hypothesis
B	0.069	Fail to reject the null hypothesis
C	0.000	Reject the null hypothesis
D	0.000	Reject the null hypothesis
E	0.000	Reject the null hypothesis
F	0.000	Reject the null hypothesis

6.8 Objective and psychophysiological measurement

Within this group of methods, the subjects' mental workload is inferred from their behaviour or performance. For completeness of the work, the objective measurement methods of the study are done using a Polar heart rate monitor. This method is a combination of the objective measurement and psychophysiological measure. When the subject felt ready for the assembly, a baseline measurement of the heart rate was performed in a sitting position. This was followed by the tasks in random order. The heart rate is measured as it is quite easy and inexpensive, using any number of readily available commercial devices. Based on the available research literature, the following hypotheses are formulated:

Hypothesis 1: *Null hypothesis*, H_0: There will be no changes in heart rate within each task.

Hypothesis 1: *Alternative hypothesis*, H_1: There will be significant change in heart rate within each task, particularly on those of higher cognitive load than those of lesser load.

Hypothesis 2: *Null hypothesis*, H_0: There will be no difference in heart rate between skilled and unskilled subjects.

Hypothesis 2: *Alternative hypothesis*, H_1: There will be a significant difference in heart rate between skilled and unskilled subjects.

A Pearson correlation was performed and statistically significant correlations for both (skilled: $r = 0.639$, $p < 0.001$; unskilled: $r = 0.783$, $p < 0.001$) are obtained. Each subject worked on each problem at their rate that is, some subject naturally worked on the problems more quickly or more slowly than others, the time spent on each of the individual activities is reported not as an absolute value but as a percentage of the total time spent completing the task. As per Cranford et al. [50] evidence for our initial assertion that heart rate can potentially be used to measure changes in cognitive load. Thus, it is appropriate to consider the data through the lens of the hypotheses outlined previously in this paper.

Hypothesis 1

In analysing the results, the observed change in heart rate within each task was smaller for the low and medium complexity tasks (tasks 2,3, and 4) than for the high complexity condition, thus confirming the hypothesis.

Hypothesis 2

It is observed that the average change in heart rate for the unskilled subjects is substantially greater than that for skilled subjects.

6.9 Subjective measurement

Following each task, the questionnaires (NASA-TLX, and ISA) were filled out. NASA-TLX measures the subjective workload of the subject in multiple dimensions (mental demand, physical demand, time demand, performance, effort, and frustration), which are combined to an overall value.

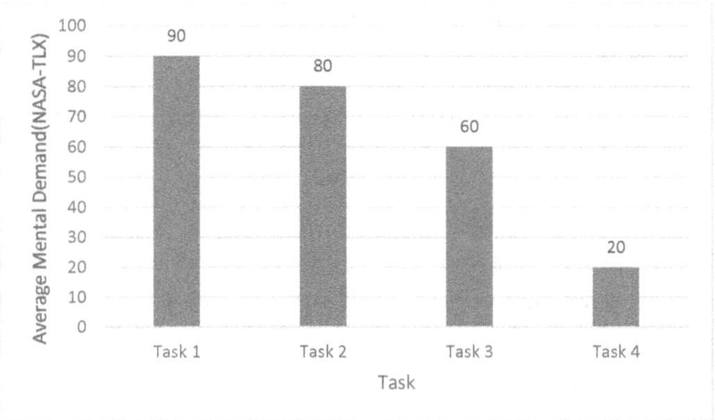

Figure 6.6 Average mental demand in each task of all the subjects

It is depicted in Figure 6.6 that the average mental demand value can decrease from task 1 to task 4 as the level of difficulty. The main effects of physical and mental demands were significant across all sub-scales and the overall workload of the NASA-TLX (all $p < 0.01$). In general, higher ratings were observed for all six scales with increased mental demands. Except for the physical demand sub-scale, all sub-scales were found to elicit higher ratings in the presence of the mental task. Additionally, the interaction between physical and mental demands was significant for two sub-scales, namely performance and effort, and for the overall workload (all $p < 0.0001$).

Instantaneous self-assessment (ISA) is a technique that has been developed as a measure of workload to provide immediate subjective ratings of work demands during the performance of primary work tasks such as air traffic control. The ISA is a self-rating method of the current mental workload during a task on a scale of 1 (low) to 5 (high). It is normally queried every 2-5 minutes during the activity. According to ISA, task 1 is at level 5 with an excessive workload with no spare capacity. So, most of the subjects will be behind on tasks, losing track of the full picture.

Task 2 is at level 4 with a high workload with very little spare capacity. The subjects reported that they could not work at this level for very long. Task 3 is at level 3 with the workload at a comfortable busy pace with some spare capacity. This task will be well in hand and busy at a stimulating pace. It could keep going continuously at this level. Task 4 is level 2, with a relaxed workload and ample spare capacity. More than enough time to complete the task. All the subjective measurements are matching with the experimental study.

6.10 Workforce planning based on cognitive abilities

Workforce Planning is the process of analysing, forecasting, and planning workforce supply and demand, assessing gaps and determining target talent management interventions to ensure that an organisation has the has the right workers possessing the right skills for the tasks assigned. Workforce planning problems vary significantly depending on the nature of the items to produce, the type of manufacturing system, the decisions to be made, and the optimisation criteria. The worker allocation in manual assembly lines usually aims at minimising production time, reducing labour costs, and ensuring the required level of quality. Cognitive load theory (CLT) is an instructional theory that focuses on how to present information and design activities to maximize learning. Here in this study the information is presented in such a way that the cognitive load of the workers must be higher so that the cognitive abilities of the workers can be captured.

In this study the pen assembly task explained in chapter 4 and virtual cognitive study in chapter 5 are combined to identify the cognitive abilities required to do the pen assembly study. The model found will also help us to shortlist the best workers. Regression models are developed using Poisson regression. The statistical test used is Omnibus Test (Test that all of the estimated coefficients are equal to zero–a test of the model as a whole. From the p-value, we can see that the model is statistically significant.) The level of significance considered for the studies are 5%. The eight cognitive abilities are used as independent variables with the total number of products assembled correctly as the dependent variable. Poisson regression is used to model count variables. It has a number of extensions useful for count models. Next, we see the Parameter Estimates. This includes the regression coefficients for each of the variables along with robust standard errors, p-values and 95% confidence intervals for the coefficients. From the p-value, we can see that the model is statistically significant. $p < 0.05$ the model is significant.

- **Personalised Model 1**

  Equal weightages were given to all the cognitive abilities and the model was found to be significant

- **Personalised Model 2**

  A combination of five cognitive abilities were combined to identify whether the model was significant.

- **Personalised Model 3**

  A combination of three abilities were combined to identify whether the model was significant.

- **Personalised Model 4**

  A combination of two abilities were combined to identify whether the model was significant.

As a result of Statistical analyses, a mapping is done between the significant cognitive abilities as shown in table 6.4.

Table 6.3 Mapping of cognitive abilities based on significance

	VP	Attention	Memory	Motorskills	Language	VSP	EF	Learning
VP			1			1	1	1
Attention								
Memory							1	
Motorskills								
Language								
VSP			1					1
EF						1		
Learning			1				1	

From this Table 6.4 it is evident that visual perception, memory, visual and spatial processing, executive functions and learning are the major cognitive abilities to be considered for this study. As per the output from model 3 the specific values from parameter estimates table is obtained. The indicator variable [Visual Perception to be 8, B value is 0.412] is the expected difference in log count between group 8 and the reference group [Visual Perception to be 9, B value is 0]. Regression coefficients indicates the expected difference in log count between a group and its reference group. Highest regression coefficient (B) is selected as the best combination while considering the groups of three cognitive abilities. Similarly, all the other combinations are considered for groups of three cognitive abilities and the values are obtained as per the table 6.4.

Table 6.4 Models and their scores of various cognitive abilities

Model	Visual Perception	Memory	Visual and spatial processing	Executive functions	Learning
A	8	5	8		
B	8	5		8	
C	8	5			9
D	6		8	6	
E	9		8		5
F	5			7	9

According to the results presented in Table 6.5, the following conditions are defined as acceptable in the performance of the workers during the tasks.

Visual Perception of a subject must be ≥ 5 | Memory score must be ≥ 5 | Visual and Spatial processing must be ≥ 8 | Executive functions score must be ≥6 | and Learning score must be ≥ 5

Based on these conditions each subject is shortlisted and depicted in the Table 6.6 below

Table 6.5 Subjects and their respective scores on selected cognitive abilities

Subject	VP	Memory	VSP	EF	Learning	Total Score
7	8	6	9	9	6	52
8	5	5	10	9	6	50
22	5	8	10	8	6	47
24	5	7	10	6	5	47
25	5	5	8	7	7	41
30	5	9	10	10	9	56

According to Table 6.6, Subject 7 has scores of 8 for visual perception, 6 for memory, and 9 for visual spatial processing, along with scores for executive functions and learning. Referring to Table 6.5, Model A is applicable to subjects who have scores of at least 8, 5, and 8 for visual perception, memory, and visual spatial processing, respectively. Therefore, Model A is applicable to Subject 7. Similarly, Model F is suitable for Subject 30.

According to the statistical analysis output 7, 8, 22, 24, 25, and 30 are the best subjects. The methodology for worker allocation is shown in figure 6.7 to identify the allocation of these subjects 7, 8, 22, 24, 25, and 30 either to Quality control or Assembler. To summarise, the table below helps to identify the best workers for Quality control and Assembler based on their cognitive abilities.

Table 6.6 Subjects and their job assigned based on selected cognitive abilities

Subject	Job Assigned
7,8,22,24,25,30	Assembler
7,8,	Quality Control

In Tasks 1 and 2, the parts of different products are mixed, jumbled, and unlabelled. Only visual perception through digital instructions and photographs is provided. When assembling a product, the worker's memory should help identify the parts used simultaneously during the assembly process. Visual and spatial processing skills assist in matching parts to the correct product. Strong executive functions are crucial for planning and assembling the two sets of six products effectively. Learning improves through trial and error during the task. All five cognitive abilities are significant, and the total score helps to assess and categorize quality control workers. Personalized Model 2 is appropriate in this context. The total number of products correctly assembled in Tasks 1 and 2 indicates the worker's potential effectiveness in quality control.

In Tasks 3 and 4, a systematic racking concept is adopted where products are grouped, reducing task complexity and making assembly easier for the worker. While a combination of all five cognitive abilities is not required, each ability individually significantly impacts performance. In such cases, personalised models 3 and 4 contribute to performing these tasks effectively. The total number of products assembled correctly in Tasks 3 and 4 indicates whether the worker is suitable as an assembler. Therefore, for quality control (Tasks 1 and 2), Personalized Model 2 is appropriate for better results, while for assembly (Tasks 3 and 4), Personalized Models 3 and 4 are more suitable.

As a strict allocation of work to quality control and assembler only six subjects (7,8,22,24,25,30) are eligible to be selected. In real life situations this won't, be possible as the employer will have to incur huge losses. In order to solve this issue some relaxations individually without considering the cognitive skills also is recommended and is depicted in figure 6.7.

As per the flowchart in figure 6.7, according to the required First pass yield to maintain the customer demand, the number of errors can be decided. The total number of products assembled correctly is considered based on the mean value. When a cognitive skill has a good mean value, and also the two conditions of product assembly tasks are fulfilled, then the workers can be allocated. Those who are not eligible can be provided training. This flowchart is a personalised model for a particular workstation in which the values can be varied depending on the productivity required. The general allotment based on the flowchart is represented as per the figure 6.8. Out of all the 30 workers, the 29th worker requires special training to improve his skills.

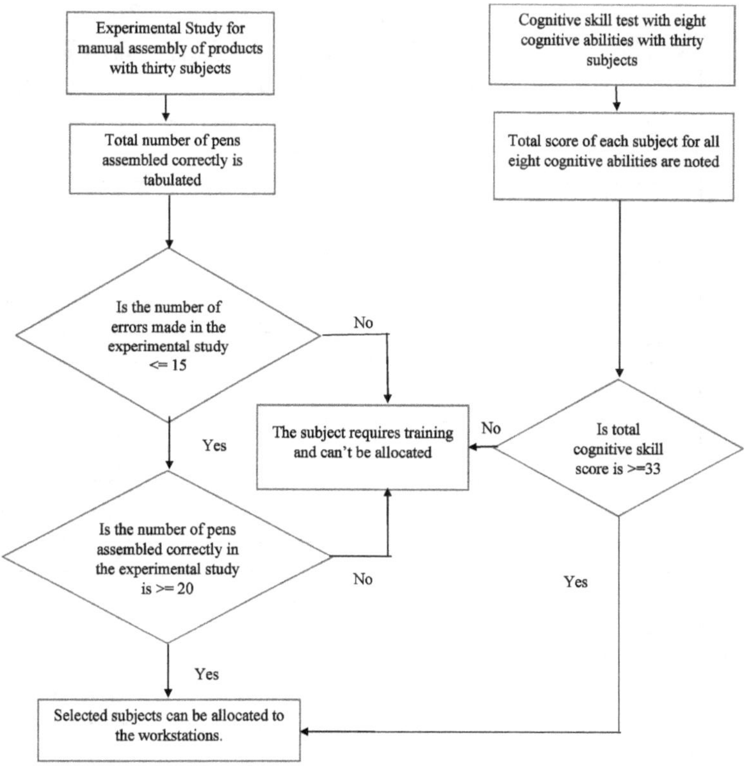

Figure 6.7 Flowchart of the methodology followed in worker allocation

Table 6.7 Pens assembled, errors in pen assembly task and the cognitive skill scores from virtual cognitive study

Subject	Total pens assembled correctly ≥ 20	Pen assembly Error ≤ 15	Cognitive ability score in virtual cognitive task (80) $\geq$33
1	26	13	45
2	28	7	44
3	29	25	41
4	23	16	39
5	5	11	32
6	9	11	48

Table 6.7 (Contd.)

Subject	Total pens assembled correctly ≥ 20	Pen assembly Error ≤ 15	Cognitive ability score in virtual cognitive task (80) ≥ 33
7	23	13	52
8	46	6	50
9	40	12	44
10	19	11	42
11	20	12	46
12	15	15	34
13	46	12	46
14	25	13	41
15	12	8	50
16	21	8	34
17	26	8	48
18	43	15	43
19	36	9	39
20	28	13	44
21	47	11	43
22	26	18	47
23	21	24	36
24	26	17	47
25	29	22	41
26	36	23	39
27	38	23	38
28	22	21	43
29	13	16	38
30	30	21	56

- Quality Control (if Error<=15)

1 2 5 6 7 8 9 10 11 12 13
14 15 16 17 18 19 20 21

- Assembler (if total pens assembled correctly>=20)

1 2 3 4 7 8 9 11 13 14 16
17 18 19 20 21 22 23 24 25 26 27
28 30

Figure 6.8 Workers allocated as assemblers and quality control

From table 6.8 from the pen assembly study, we selected the Quality control and Assembler people separately and is shown in figure 6. From the virtual cognitive task, the subjects selected will be 8,2,15,16,17,19,5,6,10,21,9,11,13,1,7,14,20,12,18. Comparing this with figure 6.8 we get the allocation considering output from Pen assembly experiment and virtual cognitive task.

Quality control – 8, 2, 15,16,17,19, 5, 6, 10, 21, 9,11, 13, 1, 7, 14, 20,12, 18

Assembler – 8, 2,16, 17, 19, 21, 9, 11, 13, 1, 7, 14, 20, 18

There are fourteen subjects who can work both as Quality control and assembler. There are five subjects which can be good in quality control and they are 15, 5,6,10,12. There are eleven workers who are not eligible when we give certain relaxations. The recommendation is to do allocation based on regression models which are developed using Poisson regression.

The subjects 22,24,25,30 is included in the poison regression where total pens assembled correctly ≥ 20 but the error is ≥ 15.

The total number of products assembled correctly in task 1 and task 2 determines whether the subject can be good at Quality control. This means that when it comes to high and medium complexity conditions if the worker can disassemble and rework without any errors. The total number of products assembled correctly in tasks 3 and task 4 determines whether the subject can be good as an Assembler. Task 3 and 4 signifies that when it comes to assembly, only with proper information processing and learning the errors in the assembly can be reduced. This study proposes a worker allocation methodology based on worker performance and cognitive skills. A subject characterization for Quality control and Assembler workstations is done in four levels with class intervals. The workers are listed based on the number of products assembled correctly in ascending order.

A cross-trained worker is a worker able to perform multiple tasks in various locations of a manufacturing system when needed. Cross-trained workers are specialized in specific tasks but also trained to perform other tasks in case of an unplanned necessity. These advantages of cross-trained workers can be utilized if we identify such workers. From this study, the framework gives the strategy to identify the cross-trained workers as 8, 2,16, 17, 19, 21, 9, 11, 13, 1, 7, 14, 20, 18 with subject characterization.

6.11 Workforce Planning methodology – Real time

Step 1: Identify the workstation in which the cognitive load is more due to a wide variety of parts and the variability in the products being assembled. Such workstations are easily identifiable by their high rates of rejections and rework, making the task of identifying them straightforward. Then select some workers for the study, with a mix of newly joined workers, workers working in the manual assembly area, and workers from other workstations. Identify the possible solutions and narrow them down. As per this study is concerned, an immediate workstation is to be added for quality control as per the company protocol for at least six months to ensure the products are error-free. This study ensures the participation of all the workers currently in the assembly station, new workers who joined, and also the employees who reach the particular workstation during their six-month rotation. The workers are mentioned as subjects in the experimental study.

Step 2: The problem can be solved using the cognitive load, cognition, and workforce reconfiguration concepts. Product assembly tasks and virtual cognitive tasks were designed to obtain a scientific solution to this industrial problem. The product assembly task was conducted by framing certain tasks similar to the actual workstation in the assembly. If the same task is provided as per the workstation, it might affect the performance as per the cognitive load theory. In case of difficulty in manipulating cognitive load across different conditions, the same workstation tasks can be given to avoid errors and rework.

A virtual cognitive task is an approach in which a cognitive ability test is conducted using an online test. Hypothesis testing was conducted for the product assembly test and the Wilcoxon test for the virtual cognitive task. Heart rate variability with certain hypotheses was also analysed. ISA and NASA-TLX subjective measurements were applied to the product assembly task to ensure that the cognitive load matches the load in the actual assembly workstation. Here two workstations are considered for an assembler and quality control as per the requirements of the company. Allocation of workers to these two workstations is also done to ensure the products are without rework and error-free.

Step 3: The data obtained as per the study identifies the right skilled person for the two workstations, which is the root cause of the whole problem. The tasks are designed in a way that will be easier for the workers who will be newly working in a particular workstation. A set of data before the implementation of the methodology adapted was compared with the data obtained after allocating the new workers as per the study, and the results justified and confirmed the relevance and importance of the study. The identification of the problem, framing of the tasks and the whole methodology is unique and can be varied according to the scenarios. The Cognitive Load Assessment for Manufacturing (CLAM) factors are easier to define such experimental studies for workers to enhance their productivity.

6.12 Summary of Chapter 6

This study aimed to investigate how well the concepts of cognitive load, cognitive abilities, and workforce planning can be combined to solve a real-time industry problem. The technique is a personalized model, and it varies on the industrial environment and other factors like the size of the industry, workers, workstations, etc.

The product assembly task was framed according to tasks similar to the actual workstation, and through certain hypotheses and statistical analysis, the results give the performance of the worker. A detailed study of the work environment was conducted before designing the tasks, and this database helps to compare it with the methodology designed in this study. A virtual cognitive task was done to identify the cognitive abilities of the worker, and thereby his/her performance can be identified. A combined methodology with the two studies provides workforce planning.

The results highlight that finding significant skills across individuals helps to identify the cross trained workers and the ones who needs training and depending upon the complexity of task and work allocation of workers can be done which improves the productivity. The methodology addressed in this study helps to measure cognitive load accurately which gives real-time results.

In future work, a collaborative approach of machine and workforce reconfigurations in a reconfigurable manufacturing system can be considered instead of manual assembly tasks. There are many theories representing cognitive load, but personalized models that allow how cognitive load manifests itself in performance variable which differs from person to person is limited.

The application of cognitive ergonomics in the production industry is time-consuming and lot of industries are not ready to apply these concepts due to the extra expenses. The joint approach with cognitive and physical ergonomics supports the worker and the employer to gain the best output. Physical ergonomics is being taken care but a cognitive ergonomic study considering the subjective measures and worker feedback with some performance studies is limited. The pen assembly task is carried out again with the same workers using wireless EEG headset. The investigation of pen assembly worker performance using this device is discussed in the next study.

Chapter 7

Investigation of pen assembly worker performance using wireless EEG headset

7.1 Introduction

A large number of product varieties, increases the variability in number of parts and their sub parts. Their shape, size, colour, texture etc will also be different which leads to cognitive load in the workers. For the worker to complete a certain task, based on the information provided, it is essential to identify the parts correctly so a proper information and material presentation is to be provided to avoid errors. In the previous chapters subjective and performance-based methods are used and the only physiological measure used was heart rate measurement. The cognitive load is measured using a wireless Electroencephalogram (EEG) headset which is a physiological method. In the past, the usage of EEG was limited to medical applications. However, given the growing interest among researchers in exploring multiple domains, wireless EEG headsets can now also be found in applications such as psychological research, manual assembly operations in industrial settings, automotive and rail industries etc. EEG signals are used to measure the electrical activity in the brain. These signals consist of different frequency components that reflect different cognitive processes in the brain. The power, amplitude, and synchronization of these frequency components can be used to determine cognitive load. EEG data is typically divided into different frequency bands, including Delta (0-4 Hz), Theta (4-8 Hz), Alpha (8-13 Hz), Beta (13-30 Hz), and Gamma (30-100 Hz). Each band reflects different cognitive processes. Emotiv Epoc X – 14 channel wireless EEG headset is used in this study. Fouad and Labib [56] explained how this wireless EEG headset can be used to acquire data. Music stimuli recognition from EEG Signals using the same wireless EEG headset is done by Suto and Jambor [109]. The outputs from wireless EEG headset helps to determine cognitive load of each scenario of the pen assembly task based on frequency bands. EEG data is typically divided into different frequency bands and reflects different cognitive processes are as follows:

1. Delta (0-4 Hz): Delta waves are the slowest-frequency brainwaves and are generally associated with deep sleep and unconsciousness.

2. Theta (4-8 Hz): Theta waves are typically associated with increased creativity, right-brain thinking, and deep meditation. They can also be related to cognitive tasks such as attention, memory, and emotion processing.

3. Alpha (8-13 Hz): Alpha waves occur during relaxation, daydreaming, and meditation. Alpha frequency is lowered in cognitive tasks that require sustained attention and increased cognitive load.

4. Beta (13-30 Hz): Beta waves occur during active thinking and concentration and are associated with alertness and cognitive processing. Beta waves have been linked to working memory, engagement, focused attention, and multitasking.

5. Gamma (30-100 Hz): Gamma waves are the fastest frequency and typically occur during higher cognitive processes such as perception, problem-solving, and decision-making.

The brain activity recorded based on these waves is considered to be a physiological method of cognitive load measurement. Performing a spectral analysis on the measured electric potential differences (by means of a fast-Fourier transform, FFT) allows to analyse the power of different frequency bands that are present in the signal. An increase in cognitive load can be measured by an increase in brain activity, i.e., oscillations within a certain frequency band with a larger amplitude as per Antonenko et al. [42] is an underlying hypothesis identified. Wireless EEG headset can be used to determine cognitive load by capturing the electrical activity in the brain based on frequency bands. The cognitive load of different scenarios of pen assembly task (study 1) is captured by repeating the experiment with wireless EEG headset. A comparison with the results of pen assembly task and the pen assembly task with wireless EEG headset is done to identify the cognitive load patterns of scenarios. The pen assembly task done with wireless EEG headset in the lab environment can be used in an industrial environment to identify the workstations with highest cognitive load. So, the studies done using the subjects can be applied to the workers in industrial environment for reducing the cognitive load of the workers.

7.2 METHODOLOGY

7.2.1 Participants

A total of thirty subjects were selected for the study, within the age range of 20 to 30 years. The subjects wear the EEG device on head and calibration is done. The subjects were asked to mentally perform the manual assembly task in 2 minutes of time duration with restricted body movement as the EEG signals are susceptible to motion artifacts which may results in noisy data as identified by Antonenko et al. [42]. The experimental setup for each scenario were arranged and given to the subject in a random order to remove the bias towards any scenario. The EEG signals were recorded while the subject performed assembly tasks and after each scenario, the subjective assessment was carried out using NASA-TLX.

7.2.2 Experimental setup

Emotiv Epoc X is a wireless EEG device (figure 7.1) used for capturing the EEG signals which is a neuro-signal data acquisition device. The setup includes Emotiv Epoc X - wireless EEG device, EmotivPRO software (laptop/PC), USB dongle and contact lens solution for electrodes.

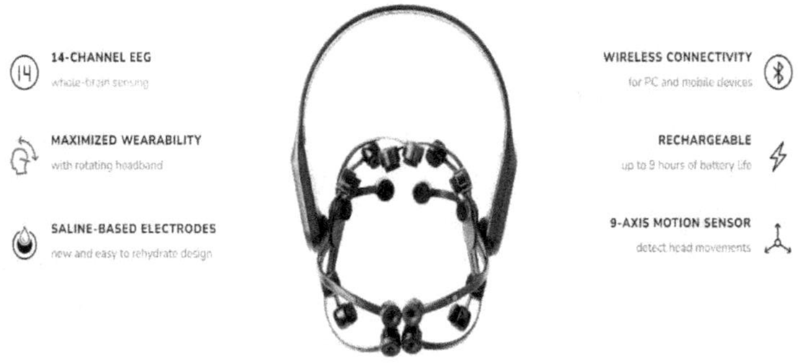

Figure 7.1 Emotiv Epoc X – 14 channel wireless EEG headset and specifications

The Emotiv Epoc X wireless EEG headset measures cognitive load by monitoring and analyzing the electrical activity in the brain. When a person is engaged in a task, their brain produces different patterns of electrical activity which is different from the patterns at rest. The level of activity in certain areas of the brain that are associated with mental effort and cognitive load can be identified. The device uses sensors placed on the scalp to detect and record these electrical signals, which are then processed by the device's software to provide insights into the subject's cognitive load. A pictorial representation of the layout of wireless EEG headset and the sensors mapping, i.e., the locations of the sensors and the references are shown in Figure 7.2.

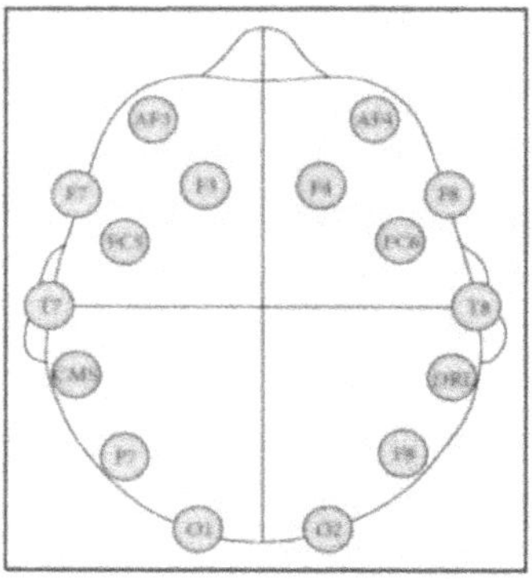

Figure 7.2 Acquiring electrodes positions on the brain according to the 10-20 model

The device has 14 channels for acquiring EEG data that are located at AF3, F7, F3, FC5, T7, P7, O1, O2, P8, T8, FC6, F4, F8, AF4 positions as per the International 10 – 20 systems. Left hemisphere of the brain is reserving the odd numbers of electrodes; even numbers of electrodes are reserved for brain right hemisphere. The electrodes F3, F4, AF3, AF4, F7 and F8 are used for neural activity imaging in the frontal lobe of the subject's brain. Electrodes T7, T8, FC5 and FC6, are imaging the temporal lobe of the brain. The parietal lobe is scanned by P8 and P7 electrodes. Neural activity of the occipital lobe is acquired with use of the O2 and O1 electrodes. The device has two references i.e., CMS/DRL references that are placed at P3/P4 or left/right mastoid process alternatively, and it comes with saline-soaked felt pads sensor material which must be moistened using a saline solution each time before use. Common Mode Sense (CMS) on the left side and Driven Right Leg (DRL) on the right side are two referencing electrodes used for signal noise reduction. Signals are sampled with sampling rate equal to 128 Hz and sent to the dongle through the wireless connection, which is connected to the USB port of personal computer and used as a receiver. EPOC X headset is connected to the computer/laptop via Bluetooth, then EEG can only run at 128 Hz and motion data can only run at 64 Hz.

7.2.3 Experimental Procedure

The EEG signals of subjects with varying age and gender were recorded according to the factorial experiment design of the pen assembly experiment.

1. Connecting and fitting the headset

The headset will be connected to the dongle in the computer's USB port. The connection will be indicated in the EmotivPRO software. The headset is placed on the subject's head according to the instructions given on the computer screen as shown in figure 7.3. The device fitting, contact quality and EEG quality has to be checked and follow the instructions accordingly.

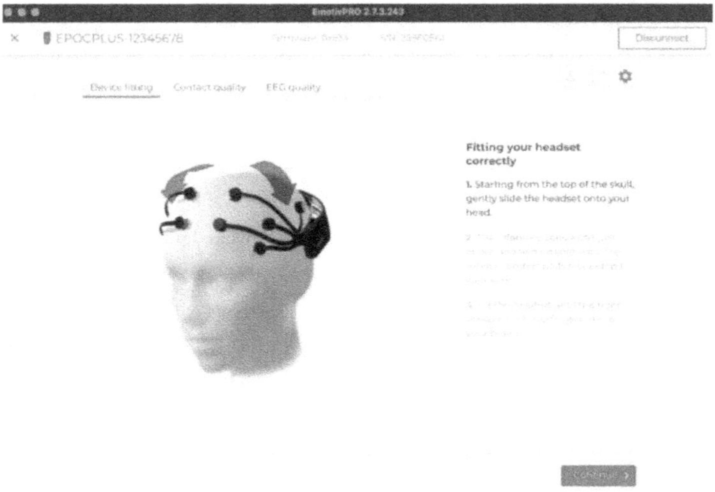

Figure 7.3 Screenshot of the Emotiv PRO software interface and instructions

Accurate EEG data collection depends on good sensor contact with the scalp and good EEG signal quality. The contact quality map shows a visual representation of the current contact quality at each individual sensor on headset. The contact quality status will be 0% as per figure 7.4 initially. Each sensor's contact quality status can be seen in real time and can be adjusted to achieve 100% contact quality.

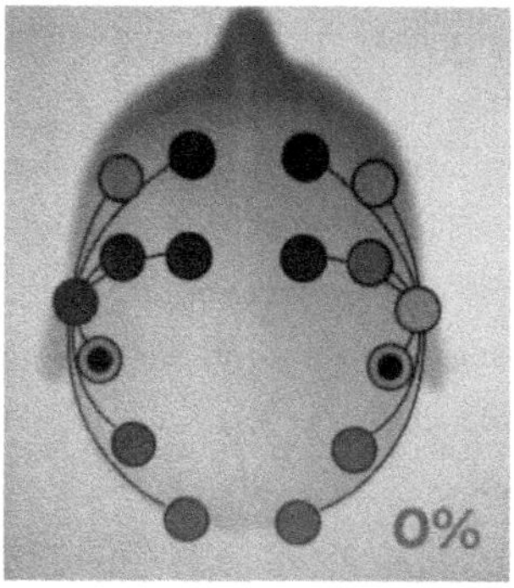

Figure 7.4 Screenshot of Electrodes in EmotivPRO software during calibration

EEG Quality (EQ) has been built to give an easy way to assess whether recording high quality brain signals or not. Basically, it is a scoring system for the quality of the signal in each sensor. Each sensor is given a black, red, orange, light green or dark green color to indicate the quality of the EEG signal for that specific sensor.

2. Data Acquisition

EmotivPRO software allows to view data streams (figure 7.5) and frequency analyses in real time whenever the headset is connected. EmotivPRO also allows to open and playback saved recordings, including all data streams and frequency analyses. The data acquired will be saved in the cloud.

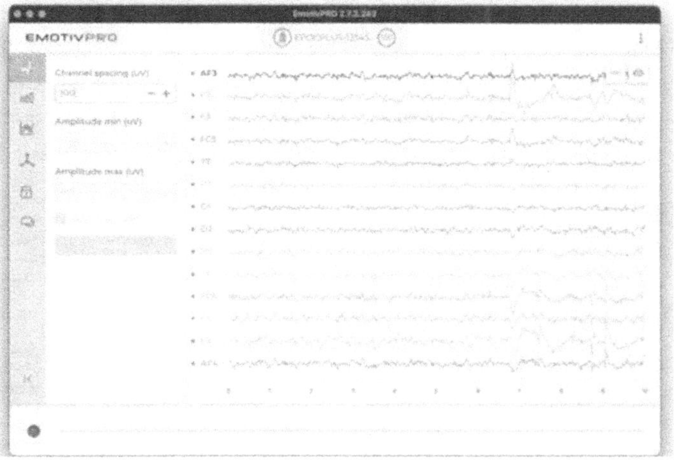

Figure 7.5 Live data stream of EEG signals in Emotiv PRO software

EmotivPRO allows to take recordings of all data streams for playback, analysis, and export. The EEG data streams can be viewed while recording. There is a baseline protocol of 40 seconds before the start of experiment by opening and closing the eyes. The data saved in cloud can be downloaded as a csv file and analysis can be done.

3. Analysis of Data collected

The exported CSV data (figure 7.6) from the software contains various data from which average of POW. <sensor>. <band> is considered. POW represents the power within a specific frequency band (band power).

POW. <sensor>. <band> – (14 channel *5 bands) 70 combinations data will be obtained from the csv file. The bands are: Theta, Alpha, BetaL, BetaH and Gamma in Emotiv epoc wireless EEG headset. For example, for AF3, the file has 5 columns:

POW.AF3.Theta, POW.AF3.Alpha, POW.AF3.BetaL, POW.AF3.BetaH, POW.AF3.Gamma. Each subject's data from all 8 scenarios are merged into a single CSV and the scenario numbers are labelled (i.e., a scenario column has been added to the merged data which specifies the scenario number of the corresponding data) A pivot table consisting of the average of POW. <sensor>. <band> intensity is created (here "average" is taken as the measure of central tendency instead of the median and the mode because of the huge data points)

From the average of POW. <sensor>. <band> for all the scenarios, the scenario is noted whose average band power intensity is greater compared to other scenarios, and this trend is observed across all the bands Power of the EEG data - The values are absolute, the unit is uV^2 / Hz (micro-Volts squared per Hertz). As per the pen assembly setup with wireless EEG headset (figure 7.7) there will be eight csv files for each subject.

Time	POW.AF3.Theta	POW.AF3.Alpha	POW.AF3.BetaL	POW.AF3.BetaH	POW.AF3.Gamma
0.07027	0.952011	0.898607	0.242765	0.129872	0.052997
0.19521	0.481214	0.745238	0.274515	0.152406	0.048068
0.32014	0.471467	0.652254	0.310824	0.171158	0.046538
0.44507	0.508419	0.62249	0.340886	0.1801	0.048434
0.57001	0.552556	0.636654	0.366031	0.175668	0.052348
0.69494	0.646566	0.679458	0.403125	0.159539	0.056688
0.81987	0.838603	0.740945	0.476877	0.138625	0.060168
0.9448	1.262587	0.851652	0.59532	0.122424	0.062021
1.06974	2.071209	1.055785	0.736467	0.117473	0.062375
1.19467	3.284051	1.353696	0.850678	0.123976	0.061764
1.3196	4.643034	1.689786	0.890933	0.137731	0.060454
1.44454	5.790958	1.970788	0.844006	0.152646	0.058674
1.56947	6.364711	2.108876	0.742068	0.165299	0.056965
1.6944	6.69751	2.133054	0.63112	0.17338	0.05591
1.81934	7.526492	2.187725	0.562616	0.175399	0.056143
1.94427	9.452452	2.407355	0.554906	0.17135	0.058023
2.0692	12.398844	2.822961	0.590216	0.161286	0.061763
2.19414	15.525002	3.331005	0.627999	0.146566	0.067274

Figure 7.6 Screenshot of the CSV file data

Figure 7.7 Subject wearing wireless EEG headset and performing the task

7.3 RESULTS AND DISCUSSION

7.3.1 Pen assembly study with 30 subjects

The pen assembly task in chapter 1 is for 15 subjects, but this study is done for thirty subjects. The data obtained is depicted in table 7.1. The results reveal that subjects have assembled a maximum number of pens in scenarios 6 and 8, i.e., a scenario where the subjects are provided with part numbered boxes with instructions or photograph. However, there exist variations in the minimum number of pens assembled. This variation is due to the change in their skillset. The success rate of the tasks is very low because the subjects are not experienced in these tasks earlier. Even though the success rates are low, chances of a learning effect are seen from the lower number of pens assembled in scenario 1 to a greater number of pens in scenario 8.

As per table 7.2 Scenario 8 (part numbered box with photograph) has the highest number of pens assembled correctly. Scenario 2 (grouped parts with instructions) has the lowest number of pens assembled. The error count was higher for scenario 7(separated parts in different boxes with photograph) and lower for scenario 6(part numbered box with instructions) according to Table 7.3. The error counts are higher for these scenarios because more attempts were made in these scenarios.

Table 7.1 Pen assembly study output of 30 subjects

Subject	Total	Max	Min	Max pens assembled - scenarios	Min pens assembled - scenarios	Success Rate
1	26	11	0	Scenario 6	Scenario 1,2,5	0.16666666
2	28	15	0	Scenario 8	Scenario 1,2,5	0.17948717
3	29	11	0	Scenario 8	Scenario 5	0.18589743
4	23	10	0	Scenario 6 & 8	Scenario 1,2,4,5	0.14743589
5	5	5	0	Scenario 6	Scenario 1,2,3,4,5,7,8	0.03205128
6	9	6	0	Scenario 8	Scenario 1,3,4,5,6	0.05769230
7	23	13	0	Scenario 8	Scenario 1,2,5	0.14743589
8	46	12	1	Scenario 8	Scenario 1,2,5	0.29487179
9	40	13	0	Scenario 8	Scenario 1,2	0.25641025
10	19	5	0	Scenario 4 & 8	Scenario 1,2,5	0.12179487
11	20	6	0	Scenario 6 & 8	Scenario 2,5	0.12820512
12	15	5	0	Scenario 6	Scenario 1,2,5	0.09615384
13	46	17	0	Scenario 8	Scenario 2	0.29487179
14	25	12	0	Scenario 8	Scenario 1,2,5	0.16025641
15	12	5	0	Scenario 6	Scenario 1,2,4,5	0.07692307
16	21	7	0	Scenario 6	Scenario 1,5	0.13461538
17	26	11	0	Scenario 8	Scenario 1,2,5,7	0.16666666
18	43	13	0	Scenario 6	Scenario 2	0.27564102
19	36	11	0	Scenario 8	Scenario 2,5	0.23076923
20	28	13	0	Scenario 8	Scenario 2	0.17948717
21	47	13	0	Scenario 8	Scenario 1	0.30128205
22	26	8	0	Scenario 6	Scenario 2	0.16666666
23	21	10	0	Scenario 8	Scenario 1, 2	0.13461538
24	26	10	0	Scenario 8	Scenario 1,2,5	0.16666666
25	29	14	0	Scenario 8	Scenario 1,2,4,5	0.18589743
26	36	14	0	Scenario 8	Scenario 2,3,5	0.23076923
27	38	19	0	Scenario 8	Scenario 1,2,3,5,7	0.24358974
28	22	6	1	Scenario 6	Scenario 1,2,5	0.14102564
29	13	4	0	Scenario 6	Scenario 2,5	0.08333333
30	30	12	0	Scenario 6	Scenario 1,2	0.19230769

Table 7.2 Total number of pens assembled correctly for all 30 subjects as per the scenario

	Intermixed	Grouped	Separated parts in different boxes	Part Numbered box
Instructions	18	7	13	233
Photograph	68	93	91	285

Table 7.3 Total no of pens assembled and the total errors made for eight scenarios

Scenario	Total Pens assembled correctly by the 30 subjects	Sum of the total errors made by the 30 subjects
Scenario 1	18	39
Scenario 2	7	52
Scenario 3	68	66
Scenario 4	93	79
Scenario 5	13	58
Scenario 6	233	14
Scenario 7	91	94
Scenario 8	285	32

Scenario 6 and 8 combined are best suited for subjects who are weak with numbers, but for an average subject there is an information overload which might increase the cognitive load. Scenario 7 has higher number of errors because the subject tries to assemble it easily with the photograph but the instructions which has the details about specific parts is not considered. The instructions will have specific details about the number of parts and the type of spring used in the pen. This information will be absent and all the subjects will put wrong springs which increased the number of errors. The next lowest error count is for scenario 8 which also suggests part numbered box with photograph. The variation in the number of pens assembled and the total errors made by thirty subjects, confirms that there is an influence of information and material presentation on the performance. A performance-based method is used here and huge margin of a greater number of pens assembled correctly is for Scenario 6 and 8. Evaluating outcomes, such as error gives an idea to infer the cognitive load. Scenarios can be written in the ascending order based on the number of errors as Scenario 6,8,1,2,5,3,4,7. For the total number of pens assembled in the whole study the order will be based on the descending order Scenario 8,6,4,7,3,1,5,2.

7.3.2 Pen assembly study with EMOTIV EPOC X

The pen assembly task is done with the subject wearing the wireless EEG headset. As per the pen assembly setup with eight scenarios there will be eight csv files for each subject to be analyzed. EmotivPRO software performs a spectral analysis on the measured electric potential differences (by means of a fast-Fourier transform, FFT) allows to analyses the power of different frequency bands that are present in the signal. The exported CSV data from the software which comes with EEG brain wear contains various data from which frequency band data that indicates the average frequency band power for each EEG sensor of the headset is further analyzed. Each subject's data from all 8 scenarios are merged into a single CSV and the scenario numbers are labeled (i.e., a scenario column has been added to the merged data which specifies the scenario number of the corresponding data). Then using the pivot table feature, a pivot table consisting of the average of band power intensity is created (here "average" is taken as the measure of central tendency instead of the median and the mode because of the huge data points). Then from the average band power intensity for all the scenarios, the scenario is noted whose average band power intensity is greater compared to other scenarios, and this trend is observed across all the bands. After determining the scenarios whose average was greater from each of the waves, another pivot table is created which counts the number of times the average of band power intensity of different bands for some scenarios was maximum. For every subject, table 7.4 is obtained after following the above procedure and it is shown below.

The various wave column indicates the number of times for any wave the average frequency band power was higher than the other waves for all the 14 sensors, for example, 14 Alpha waves correspond to each EEG sensor, and while performing the first scenario, the average power of Alpha band for six electrodes was higher than the average power of

Alpha band while performing the other scenarios. The final table is obtained by averaging all the subject's data, and the final result is shown in table 7.5.

Table 7.4 Count of the number of times the respective band power was maximum for any given subject

Scenarios	Alpha	BetaH	BetaL	Gamma	Theta
1	5	2	0	2	7
2	3	2	0	0	2
3	1	4	4	2	1
4	1	2	1	1	1
5	2	2	2	2	2
6	0	1	1	3	0
7	1	1	2	1	0
8	1	0	4	3	1
	14	14	14	14	14

Table 7.5 Average of all the subject data from table 7.4

Scenarios	Alpha	BetaH	BetaL	Gamma	Theta
1	6	2	1	2	5
2	3	2	1	1	3
3	1	4	3	1	1
4	1	2	2	1	1
5	1	1	2	2	2
6	0	1	1	2	0
7	0	1	1	2	0
8	2	1	3	3	2

In figure 7.8 the graph shows the variation of the average of band power with different scenarios and the amplitude of Theta band power shoots up more frequently for the scenarios with higher cognitive load while some with lower cognitive load. Scenario 1,2,5 experienced higher cognitive load which is a similar pattern as they assembled the least number of pens 18,7, and 13 respectively.

Scenario 8 (Part numbered box with photograph and not instructions) is having the highest number of pens assembled and the highest cognitive load as per the graph.

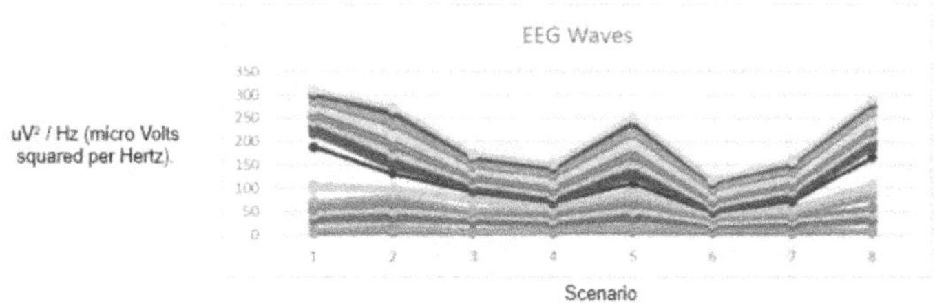

Fig. 7.8 Plot of EEG Average of Band Power vs different scenarios

7.3.3 NASA-TLX

After the subject has performed the task, its subjective load measurement is done through the NASA-TLX method. According to Prabaswari et al. [110] NASA TLX is the most widely used technique for assessing cognitive load. It has six indicators and an index of total score is provided. Rating each indicator on a scale of 0 to 100 is done by the subjects for all the eight scenarios. The final scores of each subject in different scenarios are obtained and then that average of scores is shown in table 7.6.

Table 7.6 NASA-TLX score corresponding to different scenarios

Sl. No.	Material Presentation	Information Presentation	NASA-TLX Score
1	Intermixed Parts	Instruction	80.37
2	Grouped Parts	Instruction	77
3	Intermixed Parts	Photograph	51.09
4	Grouped Parts	Photograph	45.29
5	Separated Parts in Different Boxes	Instruction	70.41
6	Part Numbered Box	Instruction	37.14
7	Separated Parts in Different Boxes	Photograph	43.05
8	Part Numbered Box	Photograph	65.12

The scenario 1(intermixed parts with instructions) resulted in a very high cognitive load whereas the scenario 6 (part numbered box with instruction) least cognitive load.

7.4 Summary of Chapter 7

There methods to measure cognitive load, used in this study are subjective methods, performance-based methods and physiological methods. In terms of subjective method, NASATLX is used. Performance based method is the error count of subjects and physiological method is the EEG method.

As per pen assembly experiment scenario 6 and 8 has least number of errors and scenario 4,7 has the highest number of errors. The number of pens assembled correctly is highest for scenario 8 and 6 and scenario 5 and 2 has the least. The experiment with wireless EEG headset revealed that the observed differences in Theta and Alpha band power among the different scenarios indicate that these average band powers are sensitive to variations in cognitive load. A higher cognitive load, was seen in scenarios 1, 2, 5, and 8 were more demanding and complex for the subjects. Scenario 6 (part numbered box with instructions) has the lowest cognitive load. This suggests that the combination of clear visual cues (part numbered box) and accompanying instructions reduced the mental effort required by the subjects. The total number of pens assembled without errors is 46.28 % whereas the number of errors made is 53.71 % considering 808 pens assembled by 30 subjects. Photograph as an information presentation give better results, but for scenario 8 photograph created confusion and the cognitive load was increased.

These findings highlight the importance of both material presentation and information presentation in workstation design which has manual assembly tasks. The way materials are organized and labelled can simplify the task and reduce cognitive load. The material and information presentation can influence the worker performance. By optimizing material and information presentation, manufacturers can potentially improve the efficiency and effectiveness of assembly processes, leading to enhanced productivity and reduced mental fatigue for workers.

Chapter 8

Effect of ergonomic aspects on single- and mixed model assembly-line balancing problems

8.1 Introduction

Ergonomics is an important aspect to consider when solving problems related to manual assembly because of its effects on productivity and ergonomics-related issues. The assembly process is characterized by frequent execution of repetitive movements and handling of components, which can potentially lead to worker overload. A suitable allocation of assembly tasks according to workers' abilities reduces the workload and creates a work environment that is ergonomic.

Businesses now understand the value of effective production systems in the face of the fierce competition between developing and changing industries [14]. In such assembly systems, lowering manufacturing costs and achieving customer demand are possible only with modifications to traditional production systems. In an assembly system, unidentified problems can create issues in finished products. For example, in the automobile industry, if an O-ring or small part is misplaced in a subassembly by a worker, the entire functionality is affected. The challenge was to identify what went wrong during the assembly process. Kosina [111] explains the Shainin methodology, which leads companies to use the automotive sector as an effective problem-solving tool. The Shainin method aids industries in assembling barrels to assemble a full diesel fuel-injection system.

Ongoing progress in the understanding of human cognitive skills requires new data acquisition, synthesis, and integration through studies and laboratories. Giridhar and Panicker [112] conducted a case study in which workers allocation is done considering cognitive abilities. Worker allocation in an assembly line was done based on cognitive aspects of information and material presentation matters. The experimental methodology analyzed the skill level of the worker and the skills required for each task, providing input for this study. This study presents a generalized version of worker allocation and verifies the objective and subjective assessments using mathematical proofs. This is a case study of a small-scale industry in southern India. This study considered the problem of 10 tasks and 6 workers for a single-product assembly line and six products for a mixed model assembly line. The inputs from a real-time industry are the precedence relationship, constraints, and objective functions for a single-product and mixed model assembly line.

In several developed countries, employees' mental health and wellbeing are addressed by identifying the appropriate skills. In developing countries, such as India, more consideration needs to be given to the right skills and workplace for workers with specific skills; instead, workers are randomly allocated. Therefore, this study formulated a mathematical model with inputs from a small-scale industrial unit to address this problem. Although many studies have focused on physical ergonomics and workstation design, cognitive aspects require attention. In manual assembly lines, mental stress decreases productivity because of the repetitive and unfavorable working conditions. Thus, cognitive skill assessment helps place them at the correct workstation. This research proposes a systematic approach to allocating workers to the workstations based on their skills.

Assembly-line balancing (ALB) is performed by assigning tasks to various workstations in the production system. These tasks involve fundamental activities to convert raw materials into finished products. Line balancing is a well-known optimization technique used in production systems. A mass-production system assembles identical or interchangeable parts or components into a final product in multiple phases at various workstations [113]. The implementation of line-balancing techniques must be refined with the advent of digital transformation. The assembly-line balancing problem (ALBP) is concerned with the optimal assignment of various assembly operations to production line workstations. The purpose of the ALBP is to optimally assign various operations in an assembly process to production line workstations. Dolgui and Proth [105] discussed the economic impact of ALBP on the design of linear production systems. Researchers have developed different algorithms and programming techniques to reduce complexity in this context. This approach is essential for solving problems for a firm to economically maintain an assembly line's productivity and flexibility. Battini et al. [114] introduced a new multi-objective model to solve ALBPs based on a recently developed predetermined motion energy system technique to rapidly estimate energy expenditure values. A real-world case study with a simple numerical example analyzed Pareto frontier behavior and identified variations in parameters related to energy and time values.

Bautista et al. [115] introduced an approach for balancing assembly lines that simultaneously considers temporal, spatial, and ergonomic factors. They conducted a case study with Nissan using several mathematical models developed to study its behavior. Assembly lines are mass-production systems that manufacture similar products in large volumes at a low cost. Manufacturers utilize mixed-model assembly lines to generate customized and nonidentical items in response to customer demands. Battini et al. [116] prevented ergonomic risks by using an integrated plan to solve the ALBP. Only a few studies have explored component picking at assembly lines and its significance in the ALPB.

This study addresses the multi-objective optimization of the ALBP, incorporates ergonomic considerations in designing production systems, and explores the applicability of the Pareto optimality concept. To this end, this study proposes an original approach based on an extension of two leading meta-heuristic algorithms: the genetic algorithm and simulated annealing. The proposed approach significantly contributes to ergonomics in assembly line research by applying Pareto optimality over a weighted approach, as reported in the literature.

The proposed approach uses the following two algorithms to solve the ALBP for single product and mixed models:

- Non-dominated sorting genetic algorithm II (NSGA-II)
- Multi-objective simulated annealing (MOSA)

Finally, this study evaluated the performance of the NSGA-II against the MOSA. This study explores the optimization of assembly line ergonomics, with a focus on real-time manufacturing units.

8.2 Modeling methodology in an ALBP

The mathematical model that formed the basis of this study was obtained from the multi-objective optimization model developed by Dalle and Dini [9]. The energy expenditure for various micromovements in an assembly line was computed using the empirical relationships in Garg et al. [94].

8.2.1 Mathematical model

The decision variables, objective functions, and problem constraints are described in the following subsections.

Notations applied

Decision variables

m Number of workstations

S_n Skill of worker assigned to n^{th} workstation and $n = 1,...,m$

η_n Energy gap of nth workstation and $n = 1,..., m$

$\bar{\eta}$ Mean energy gap between workstations

Objective functions

Objective function 1

The first objective function is to minimise the number of workstations m.

$$\text{Min } O1 = m \tag{8.1}$$

Objective function 2

The second objective function is to minimise the total skill of assigned workers.

$$\text{Min } O_2 = \sum_{n=1}^{m} S_n \qquad (8.2)$$

Objective function 3

The third objective function minimises the standard deviation of the energy expenditure of workers at workstations.

$$\text{Min } O_3 = \sqrt{\frac{\sum_{1}^{m} \bar{\eta} - \eta m}{m}}, \qquad (8.3)$$

where $\bar{\eta}$ is the mean of the energy gap between workstations and

ηm is the energy gap of workstation m and is computed using Equation (8.4).

$$\eta m = 1 - \frac{Ej}{0.0143 \times OELj \times CT}, \qquad (8.4)$$

where E_j is the sum of the energy required to perform all tasks by worker j at workstation m.

OEL_j is the occupational energy limit of worker j and

CT is cycle time (min).

Constraints: The above-stated objective functions are imposed with the following constraints:

- Precedence constraint
- Skill constraint
- Energy constraint
- Cycle-time constraint

The constraints are explained as follows.

Precedence constraint

Each assembly sequence followed a precedence constraint. The precedence relation matrix $P_{i,z}$ is given by Equation (8.5)

$$P_{i,z} = \begin{cases} 1 & if\ task\ i\ precedes\ task\ z \\ 0 & if\ task\ i\ doesnot\ precede\ task\ z \end{cases} \qquad (8.5)$$

Skill constraint

The worker assigned to the workstation should have a skill (Sj) greater than or equal to the maximum skill ($Smax_m$) required for the task performed at workstation m, which is given by Equation (8.6).

$$S_j \geq Smax_m \quad (8.6)$$

where,

$$S_i = \begin{cases} 0, & for\ low\ skill\ level \\ 1, & for\ intermediate\ skill\ level \\ 2, & for\ the\ task\ that\ requires\ high\ skill \end{cases} \quad (8.7)$$

Energy constraint

Workers should not work beyond their physical capacity. The energy required for micro-movement n of task i by worker j (e_{nij}) was computed using the empirical relation proposed by Garg et al. [94].

Some of the relations applied are as follows:

The energy required for arm lowering,

$$e_{nij} = 10^{-2}\times[0.093 \times WW_j\ (h2_n - 0.81) + (1.02 \times L_n + 0.37 \times g_j \times L_n) \times (h2_n - h1_n)] \quad (8.8)$$

The energy required for arm lifting,

$$e_{nij} = 10^{-2}\times[0.062 \times WW_j\ (h2_n - 0.81) + (3.19 \times L_n + 0.52 \times g_j \times L_n) \times (h2_n - h1_n)] \quad (8.9)$$

The energy required for walking,

$$e_{nij} = 10^{-2}\times (51 + 2.54 \times WW_j \times (SW_n)^2 + 0.379 \times WW_j \times GS \times SW_n) \times t_n \quad (8.10)$$

The total energy expenditure of worker j to perform micromotion n at workstation m, with a subset of $1 \ldots Nt$ tasks are given by (8.11):

$$E_j = \sum \sum_{n=1}^{Nmi} e_{nij}\ ,\ i = \text{subset of } 1,..,N_t \quad (8.11)$$

where Nmi is the total number of micromovements in task I and Nt is the total number of assembly tasks.

Furthermore, the Garg model equations for various micromovements are as follows:

For carrying loads held against the thighs or against the waist (in kcal)

$$e_{nij} = 10^{-2}[68 + 2.54\ WW_J \text{ x } SW_n^2 + 4.08\ L_n \text{ x } SW_n^2 + 11.4L_n + 0.379\ (1+ WW_J)\ GS \text{ x } SW_n]t_n \quad (8.12)$$

For pushing/pulling at a bench height of 0.8 meters (in kcal/push)

$e_{nij} = 10^{-2}(112\ WW_J + 1.15F + 0.505\ GS \times F)$ (8.13)

For lateral movement of the arms at 180° in both hands (kcal/lateral movement of the arms).

$e_{nij} = 10^{-2}(0.11\ WW_J + 0.726L_n)$ (8.14)

For holding at arm's length in one hand (in kcal),

$e_{nij} = 0.088\ L_n \times t_n$ (8.15)

For pushing/pulling, at a 1.5-meter height (kcal/push),

$e_{nij} = X(0.086 + 0.036F)$ (8.16)

Carrying loads held at arm length at the sides (in one or both hands) (kcal),

$e_{nij} = 10^{-2}[\,80 + 2.43\ WW_J \times SW_n^2 + 4.631 \times SW_n^2 + 4.62L + 0.379\,(1 + HW_j)\ GS \times SW_n]\ t_n$ (8.17)

For holding at arm's length, against thighs, or at sides (both hands) (kcal),

$e_{nij} = 0.037\ L_n \times t_n$ (8.18)

For holding against the waist (in kcal),

$e_{nij} = 0.062\ L_n \times t_n$ (8.19)

where

E is the rate of metabolism (in kcal min),

e_{nij} is the energy expenditure (kcal) for carrying, walking, or holding,

WW_J is the weight of the operator (in kg),

F is the average force applied to push or pull the hands (kg),

GS represents the grade of the walking surface (%),

$h1_n$ is the height from the ground to the starting point of the lifting or endpoint of the lowering micromotion,

$h2_n$ is the height from the ground to the endpoint of lifting or the starting point for lowering the micromotion,

L_n represents the weight of the load (in kg),

g_j represents the gender of the operator, namely 0 for females and 1 for males,

SW_n is the speed of walking (in m/s),

X is the horizontal movement of' work piece (in m), and

t_n is the time taken (in min).

The energy expenditure was calculated using the Garg model equations.

The physical limit of workers for professional applications is assumed to be 35% of the maximum energy expenditure rate, $Wmax_j$ [in W], thus providing the occupational energy expenditure (OEL_j) [in W]. Occupational energy expenditure is the power required for a worker to work for 8 h, which is the limit of the worker and is calculated using Equation (8.20).

$$OEL_j = 0.35 \times Wmax_j \qquad (8.20)$$

The maximum energy expenditure rate for a worker was calculated based on the maximum oxygen consumption, VO_{2j} [l/min]. The equivalent energy at maximum oxygen consumption was assumed to be 21.25 kJ/l, as per Dalle and Dini [9]. Using this value, the energy expenditure rate at maximum work is obtained using Equation (8.20).

The metabolic equivalent (MET) is a measure of the accumulated metabolic workload during an activity. The mean MET during the activity was considered to be 354.3, as the amount of physical activity was minimal in this study.

$$Wmax_j = 354.3 \times VO_{2j} \qquad (8.21)$$

The maximum oxygen consumption of a worker was calculated using the formula proposed by Schacherer et al. [117], as per Equation (8.22).

$$VO_{2j} = \begin{cases} 66.734 - 0.315\, aj - 0.678\ PFATj & \text{if the worker is male} \\ 58.904 - 0.356\, aj - 0.494\ PFATj & \text{if the worker is female} \end{cases} \qquad (8.22)$$

where a_j is the age of the worker and $PFAT_j$ represents the body fat percentage, which is estimated using the relationship proposed by Jackson-Pollock, as given in Equation (8.23).

$$PFAT_j = 1.61 \times BMI_j + 0.13 \times a_j - 12.1 \times g_j - 13.9 \qquad (8.23)$$

where BMI_j is the body mass index and g_j is the gender of the worker.

Body mass index is estimated using Equation (8.24)

$$BMI_j = \frac{WWj}{HWj^2} \qquad (8.24)$$

Thus, the calculated energy limit is considered the physical limit of the worker, and the energy expenditure of the worker should not exceed this limit for each workstation, as represented in Equation (8.25):

$$E_j \leq 0.0143 \times OEL_j \times CT \qquad (8.25)$$

where CT is cycle time (min).

In the mixed-model ALB, the product mix changes frequently in addition to the stochastic task time, thus countering a range of NP-hard problems without optimal solutions. This finding highlights the need for simpler methods. The energy expenditure of workers in the mixed-model assembly lines was calculated using the virtual average model developed by Battini et al. [118]. The weighted average energy expenditure was calculated using Eq. (8.26).

$$\text{Weighted average energy} = \sum \sum p_k \times E_{ikj} \qquad (8.26)$$

where p_k is the proportion of product k in the repeating sequence and E_{ikj} is the i th task's energy expenditure of product k for worker j assigned to the workstation. Thus, the energy time constraint for mixed-model ALB is

$$\sum \sum p_k \times E_{ikj} \leq 0.0143 \times OEL_j \times CT \qquad (8.27)$$

Cycle-time constraint

In single-product ALB, the total task time of all the tasks assigned to a workstation should not exceed the total cycle time of the assembly line, as shown in Equation (8.28).

$$T_i = \sum_{n=1}^{Nm} t_n \qquad (8.28)$$

The task time (T_i) is the sum of the execution times of all the micromovements involved in the process, as given by Equation (8.29).

$$T_i \leq \text{Cycle time } (CT) \qquad (8.29)$$

In a mixed-model ALB, multiple models are assembled simultaneously; hence, the weighted average time is considered rather than the task time. The weighted average time is the time required at a workstation to perform tasks for different products. The weighted average time at each workstation cannot exceed the required cycle time for all products. The weighted average time was estimated using Equation (8.30):

$$\text{Weighted average time} = \sum \sum p_k \times T_{ik} \qquad (8.30)$$

where p_k is the proportion of product k in the repeating sequence and

T_{ik} is the task time for the ith task of product k assigned to the workstation.

Therefore, the cycle-time constraint for the mixed-model ALB is given by Equation (8.31).

$$\sum \sum p_k \times T_{ik} \leq \text{Cycle time } (CT) \qquad (8.31)$$

8.2.2 Problem instance for a single-product assembly line

In this study, a problem consisting of ten tasks and six workers was considered. The characteristics of workers available for the assembly line and their skills are listed in Table 8.1

Table 8.1 Characteristics of workers available and their skill level for an assembly line

Worker	Weight (kg)	Age (years)	Gender	Height (m)	Skill level	
					Single product assembly line	Mixed model assembly line
1	75	50	Male	1.81	1	1
2	85	45	Male	1.7	2	2
3	55	30	Female	1.59	1	2
4	70	50	Female	1.65	2	2
5	70	30	Male	1.75	0	2
6	63	40	Female	1.7	0	1

The skill level of the assigned worker should be greater than that required to perform a task at a workstation. The skill levels required for each task in the assembly lines are presented in Table 8.2. The micro-movements involved in each task of the considered problem instance are also listed.

Table 8.2 Skill level required for each task in the assembly lines

Task		1	2	3	4	5	6	7	8	9	10
Micro-movements		Arm lowering	Walking	Arm lifting	Carrying load against thighs or waist	Arm lifting, walking, and arm lowering	Pushing at bench height	Arm lifting	Arm lifting, lateral movement of arms (180°), and arm lowering	One hand lifting, holding at an arm’ s length in one hand, and arm lowering	Pushing at 1.5-m height
Skill level required	Single product assembly line	1	0	1	0	1	1	0	0	2	0
	Mixed model assembly line	0	2	1	0	0	1	0	1	0	2

The precedence constraint is that Task 1 must be completed before Task 2 and Task 9 must be completed before Task 10. To address this problem, the daily demand for a single product was 48 units with a daily working time of 8 h. Therefore, the cycle time at each workstation was computed to be 10 min. Moreover, the energy of a worker should be greater than the sum of the energies required to perform all tasks at a workstation.

8.2.3 Problem instance for a mixed model assembly line

In this study, the number of products in the mixed model assembly line was set to six. The precedence relationship for a single-product assembly line is followed in this problem instance for a mixed model assembly line. As previously stated, the skill of the assigned worker should be greater than the skill required to perform the task at the workstation. Moreover, the worker's energy should be greater than the energy required to perform all tasks assigned to a workstation. Thus, when calculating the sum of the energies in a mixed-model assembly line, the tasks required for all six products must be considered. All tasks and their sequences were the same as those for the single-product assembly line problem instance. As mentioned above, the characteristics of the workers available for assignment to the mixed-model assembly line and their skill levels are presented in Table 1 and the skill levels required for the mixed-model assembly line is presented in Table 8.2. The daily demand and cycle time for each of the six products are listed in Table 8.3. The times required for each task for the six products in the mixed-model assembly line are presented in Table 8.4.

Table 8.3 Demand and cycle time for six products

Product	Daily Demand (units)	Required cycle time (min)	Demand proportion
A	20	24	0.2
B	20	24	0.2
C	15	32	0.15
D	20	24	0.2
E	10	48	0.1
F	15	32	0.15

Table 8.4 Task time of various products in a mixed model assembly line

Task	Product task time (in min)					
	A	B	C	D	E	F
1	1.78	0.88	1.27	1.57	1.80	1.00
2	1.57	1.68	1.60	1.05	0.90	1.98
3	0.85	1.42	1.10	1.95	1.68	1.47
4	1.95	1.40	0.55	0.68	1.35	1.10
5	1.90	1.88	1.40	1.00	1.08	1.08
6	1.25	0.55	0.65	1.42	1.87	1.07
7	0.62	2.00	1.50	0.95	1.07	1.85
8	1.52	0.72	0.63	0.70	1.90	0.72
9	0.85	0.73	1.52	0.55	0.73	0.65
10	1.25	1.12	0.70	1.40	1.20	1.97

If Tasks 1 and 3 are assigned to workstation 1, then the total weighted average time for workstation 1 is calculated as follows:

Total weighted average time = $\sum k \sum i\ (qk \times tik) = 2.765$ min (8.32)

The weighted average time at each workstation should be less than or equal to the required cycle time.

8.2.4 Relaxation of an objective for solving the ALBP

As reported in the literature, the problem of deterioration becomes more prevalent when solving a multi-objective optimization model with an increasing number of objectives. Hence, a dimensionality-reduction approach was adopted to simplify the methodology. Thus, among the three objectives considered in the present study, the first was to minimise the number of workstations which is fixed to four. Four workstations are considered based on the workstations of pen recycling unit. Certain inputs from pen assembly task helped to frame the ALBP w.r.t the number of tasks, number of products, time to assemble a certain product etc.

8.3 Solution methodology proposed for an ALBP

In the present research, to solve an ALBP considering ten tasks for single- and mixed model assembly, the following two multi-objective optimization solution approaches are proposed:

- Non-dominated sorting GA II (NSGA-II) and
- MOSA

Contrary to the weighted solution approaches reported in the literature, a Pareto optimality-based approach was applied in the present study to obtain non-dominated solutions. This is the unique contribution of this study to the scientific literature on ALBP. In this study, the following two variants of ALB were considered:

3. Single-product ALB
4. Mixed model ALB

In both scenarios, ten tasks were involved in an assembly operation to be carried out at four workstations. Hence, these ten tasks are to be assigned among the available workers in the four workstations, such that the total required skills can be minimised. In addition, the variance in the total energy expenditure among the workstations should be minimised; that is, the task should be distributed according to the worker's physical capability.

8.3.1 Solution methodology for a single-product ALBP

This subsection illustrates the steps to be followed to solve the single-product ALBP using NSGA-II and MOSA.

The skills and micromovements required for the tasks are listed in Table 2. The daily demand for the products was 48 units, and the daily working time was 8 h. Using the available data and the proposed NSGA-II or MOSA algorithm, a Pareto front was developed to select the best task and worker sequences. The steps for obtaining the solutions are as follows:

Step 1: Calculate the energy required to perform each of the ten tasks by each of the six workers using the Garg model for a single product. The energy expenditure of each worker per task was calculated and tabulated based on the Garg model equations explained earlier for each task and the task frequency.

Step 2: Calculate the workers' occupational energy limit (OEL) based on worker characteristics.

Step 3: Implementation of NSGA-II or MOSA

Step 4: Generation of an initial feasible population

Step 5: Calculation of objective function values

Step 6: Development of a Pareto Front

8.3.2 Solution methodology for a mixed model ALBP

In the mixed-model problem, six products were assembled. Hence, the assembly line is balanced so that the demand for all products is fulfilled. In this scenario, the task required to be completed for all six products was the same as that in the single-product scenario. The energy available to a worker should be greater than the sum of the energies required to perform all tasks at a workstation. When calculating the sum of the energies in the mixed-model assembly line, the tasks for all six products should be considered.

8.3.3 An illustrative example of the proposed solution approach using NSGA-II

The GA starts by generating an initial population of the required size. The fitness function was then calculated for all individuals in the initial population. Subsequently, based on the fitness value, individuals with the best fitness values survive to the next generation. Crossover and mutation operations are then applied to the initial population to generate a population for the next generation. This process was continued until a new population of the required size was formed.

In this study, a chromosome was designed to represent the task sequence and the worker assignment. An order-based crossover was applied to the task sequence to ensure feasible solutions, and a swap mutation was applied to the worker assignment on a chromosome.

8.3.3.1 Generation of the initial feasible population in NSGA-II

Several feasible chromosomes, equal to the required population size, were randomly generated. Details about the solution approach is provided. The chromosome was generated as per the flowchart shown in Figure 8.1. A sample scheme for a feasibly generated chromosome is presented in Table 8.5. Similarly, an initial feasible population of a given size was generated.

Table 8.5 Feasible chromosome from initial population

<table>
<tr><td colspan="11">CHROMOSOME 1</td></tr>
<tr><td>Task sequence</td><td>5</td><td>1</td><td>9</td><td>3</td><td>10</td><td>4</td><td>2</td><td>7</td><td>8</td><td>6</td></tr>
<tr><td>Skills required as per task</td><td>1</td><td>1</td><td>2</td><td>1</td><td>0</td><td>0</td><td>0</td><td>0</td><td>0</td><td>1</td></tr>
<tr><td>Maximum skill in workstation</td><td colspan="2">1</td><td colspan="3">2</td><td colspan="3">0</td><td colspan="2">1</td></tr>
<tr><td>Worker assignment</td><td colspan="2">1</td><td colspan="3">4</td><td colspan="3">2</td><td colspan="2">3</td></tr>
<tr><td>Available skill</td><td colspan="2">1</td><td colspan="3">2</td><td colspan="3">2</td><td colspan="2">1</td></tr>
</table>

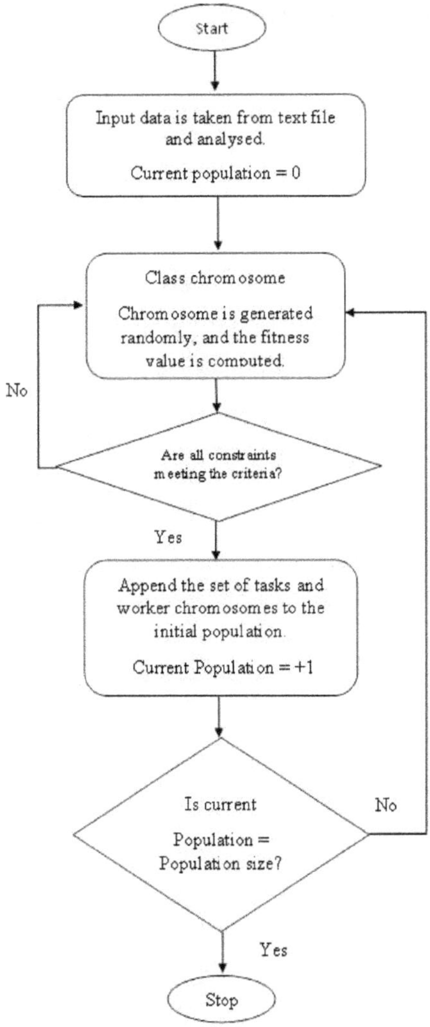

Figure 8.1 Flow chart of generating initial feasible population in NSGA-II

8.3.3.2 Calculation of objective function values

Objective 1(*O1*) minimises the number of workstations. Although the first objective was to minimise the number of workstations and reduce the complexity of computation, the first objective was relaxed in the initial phase of the work. Therefore, the number of workstations was fixed at four, and the worker assignments and task sequences were determined using NSGA-II.

Objective 2(*O2*) minimises the total required skill. For chromosome 1, the objective function value O2 is computed as follows:

$$O2 = 1+1+2+1+1 = 6$$

Objective 3(*O3*) minimised the variance in the total energy expenditure at each workstation, and the objective function value O3 was estimated as follows:

$$O3 = \sqrt{\frac{\sum_{m=1}^{wst}(\bar{n}-n_m)^2}{k}} \qquad (8.33)$$

where $\bar{n}$ is the mean energy gap between workstations.

wst is the number of workstations.

n_m is the energy gap of workstation m and is calculated using Equation (8.4)

$$n_m = 1 - \frac{Total\ E_j}{0.0143 \times OEL_j \times CT} \qquad (8.4)$$

where Total E_j is the sum of the energy required to complete a task by worker *j* at the workstation *m*,

OEL_j is the occupational energy limit of worker *j*, and *CT* is the cycle time in minutes.

8.3.3.3 Fitness calculation for all chromosomes in the feasible population

As stated above, the novel contribution of this study is the application of the Pareto optimality approach to solve an ALBP. In the Pareto optimality approach, instead of assuming weighting for the three objectives, fitness is determined based on non-dominated solutions. A Pareto front was plotted between the values of the objective functions, *O3* and *O2*, for all chromosomes in the population. In the plot, equally good chromosomes are represented as one front and assigned the same fitness value. Therefore, the chromosomes in front 1 had the highest fitness value, followed by the chromosomes in the next front.

8.3.3.4 Generation of next-generation population

Chromosomes in the next-generation population were generated based on the previous population using operators such as elitism, crossover, and mutations.

1. **Elitism:** In elitism, the given number of chromosomes with the highest fitness is directly passed on to the next-generation population to guarantee its presence.

2. **Crossover:** To satisfy the precedence constraints given for the tasks and to ensure feasible chromosomes, an order-based crossover is applied in this study. The chromosome for the crossover was selected from the population using tournament selection. The task sequence on the chromosome was subjected to an order-based crossover. As indicated in Table 8.6, an order-based crossover cuts the chromosomes of the two parents at two randomly chosen points. The first child chromosome is created using the head and tail portions of the first parent chromosome and by inserting the missing genes in that order, as they appear in the second parent chromosome. The same process, but the opposite, is performed to obtain the second child chromosome. This genetic operator ensures that both child chromosomes are feasible with respect to the precedence constraints. At each iteration, individuals in the current population were selected to produce offspring. In this study, order-based crossover was applied to the task chromosome.

As shown in Table 8.6, it works by selecting cutoff points of 2 and 8. The first offspring/child is generated by the head and tail of the first parent and the missing genes in the order in which they appear in the second parent; the opposite is true for the second offspring/child. This genetic operator was implemented to ensure that populations made only feasible solutions, which means assembly sequences respecting the precedence constraints.

Table 8.6 Example of order-based crossover applied to task chromosome

Parent 1 chromosome										
Task chromosome	4	7	1	3	2	9	6	8	10	5
Worker chromosome	4		1			3			2	

Cut points at point 2 and 8 of the task chromosome

Parent 2 chromosome										
Task chromosome	9	1	8	6	10	2	4	7	5	3
Worker chromosome	2		1			5			3	

Cut points at point 2 and 8 of the task chromosome

Child 1 chromosome										
Task chromosome	4	7	9	1	8	6	2	3	10	5
Worker chromosome	4		1			3			2	
Child 2 chromosome										
Task chromosome	9	1	4	7	2	6	8	10	5	3
Worker chromosome	2		1			5			3	

3. Mutation

A swap mutation was applied to the worker assignment in the chromosome if the mutation probability was greater than a given value Table 8.7. It is applied to a worker chromosome, and two genes are exchanged within the same individual. Specifically, this operator reverses the positions of the two workers assigned to different workstations and is implemented to ensure randomness.

This process is iterated for a given number of generations to obtain the best chromosome, comprising the best task sequence and worker assignment. For illustrative purposes, a hypothetical problem consisting of ten tasks to be performed at four workstations in both single-product and mixed model assembly line scenarios was considered.

Table 8.7 Example of swap mutation applied to worker chromosome

Parent chromosome										
Task chromosome	4	7	9	1	8	6	2	3	10	5
Worker chromosome	4		1			3			2	

Swap at point 1 and 4 of worker chromosome

Child chromosome										
Task chromosome	4	7	9	1	8	6	2	3	10	5
Worker chromosome	2		1			3			4	

8.3.4 An illustrative example of the proposed solution approach using MOSA

Simulated annealing (SA) has been reported to be a reliable approach for solving single-objective optimization problems. The simulated annealing algorithm is a metaheuristic for approximately optimizing solutions in a large search space. It is mimicked by a metallurgical process that involves heating and controlling the cooling of materials to increase their crystal size and reduce their defects. In simulated annealing, when annealed slowly and sufficiently, it converged to the global optimum. However, researchers have recently focused on optimizing multiple objectives. The simultaneous optimization of several objectives is usually impossible; thus, a curve representing two objectives or a surface representing three or more objectives describes the trade-off between objectives. These curves or surfaces are known as the Pareto fronts [119]. In the literature, simulated annealing has been adapted to solve multi-objective optimization problems by combining the objectives into a single-objective function. This is known as a weighted approach. The unique contribution of the present work is that it replaces the weighted approach and applies the Pareto optimality approach, providing a set of compromising solutions for decision-makers and adopting MOSA.

In MOSA, two sets of solution sequences were initially randomly generated. Each set contained task and worker sequences. These sequences were used as seed sequences for MOSA. In the SA algorithm, poor solutions are temporarily accepted to explore the solution space. The algorithm randomly selected a solution that was close to the current solution at each step. Its quality is then measured, which determines whether to proceed with the solution or stick to it. The advantage of the simulated annealing methods is that they require a reasonable amount of time to obtain near-optimal solutions.

8.3.4.1 Phase 1 of MOSA

In this phase, the first set of randomly generated solution sequences was used as the initial sequence for all search directions. The search direction was varied using two variables, x_1 and x_2 (relative to the total skill level required and the variance in energy expenditure).

In the first step, $x_1 = 0$, $x_2 = 1$, and $S = S_1$. Then, generate a uniform random number, u, in the range [0, 1]; if u is less than x_2, then perturbation is employed on S, and a new sequence S' is obtained. Then, this new sequence S' (obtained by the perturbation scheme) and S are compared with respect to the energy variance objective; otherwise, (i.e., $u > x_2$), S and S' are compared to Objective 2, representing the total skill level required.

The simulated-annealing, technique-based search was performed for a predefined number of iterations, and then the search process was restarted with $x_2 = x_2 - 0.1666$, $x_1 = 1 - x_2$, and $S = S_1$. This process is continued until $x_2 <= 0.5$. This search method attempts to determine as many Pareto-optimal solutions as possible within a sequence of search directions. The initial value of the direction parameter $x_1 = 0$ (or $x_2 = 0$, depending on the phase) is increased by 0.1666 in subsequent steps to divide the range [0, 1] into six regions such that eight combinations of x_1 and x_2 are obtained and used to discover Pareto-optimal solutions in the search process.

8.3.4.2 Phase 2 of MOSA

In the second phase, a second set of randomly generated solution sequences is given as the initial sequence for all search directions. The search direction was varied using two variables, x_1, and x_2 (relative to the total skill level required and the variance in energy expenditure).

In the first step, $x_1 = 1$, $x_2 = 0$, and $S = S_2$. Then, generate a uniform random number, u, in the range [0, 1]; if u is less than x_1, then perturbation is employed on S, and a new sequence *S'* is obtained. Then, this new sequence *S'* (obtained by the perturbation scheme) and *S* are compared with respect to the total skill objective; otherwise (i.e., $u > x_1$), *S* and *S'* are compared with respect to Objective 3, representing the variance in energy expenditure.

The simulated-annealing, technique-based search was carried out for a predefined number of iterations, and then the search process was restarted with $x_1 = x_1 - 0.1666$; $x_2 = 1 - x_1$; and $S = S_1$, and this process was continued until $x_1 <= 0.5$. In such a sequence of search directions, the search process attempts to uncover several Pareto-optimal solutions.

The parameters used for the simulated annealing technique were as follows:

- Acceptance-probability function, *F*
- Epoch length (number of iterations at a given temperature)
- Initial temperature, *T*
- Final temperature
- Cooling rate and
- Termination criterion

The acceptance-probability function for a new sequence S' is given by Equation (8.34).

$$F(S,S',T) = \exp\{-(\ (Z(S') - Z(S)\ ) \times 100\ /\ Z(S)\ \ )\ /\ T\} \tag{8.34}$$

In the above equation, *S'* is the sequence obtained through the perturbation scheme, called the random-insertion perturbation scheme (RIPS), employed on *S* (the initial sequence), and *T* denotes temperature. The temperature progressively decreased from the initial positive value to the final temperature. At each step, the SA heuristic randomly selects a solution *S'* that is close to the current one. If the movement to *S'* is not economical, it is decided according to the temperature-dependent probabilities of selecting better or worse solutions. The parameters were selected such that the system ultimately moved to a lower energy level. The initial temperature was 575 °C, the cooling rate was 0.9, and the final temperature was 20. The total number of iterations was 30; therefore, 30 new sequences were generated for each temperature. An illustration of this algorithm is shown in Figure 8.2.

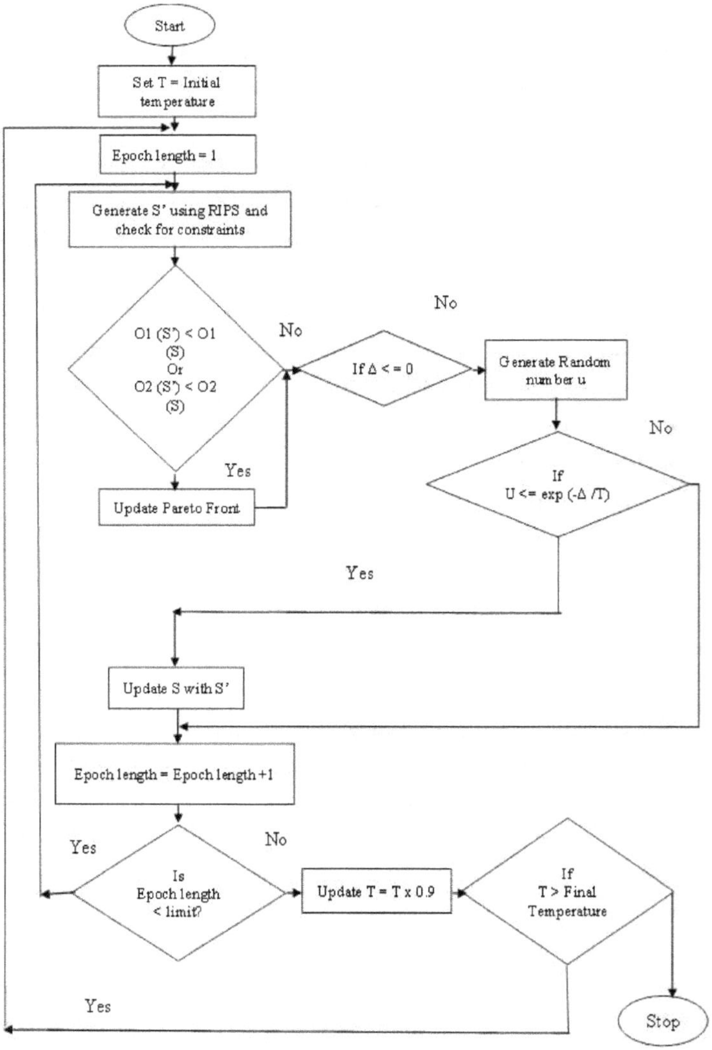

Figure 8.2 Flowchart of MOSA approach used for plotting Pareto front

8.4 Results

An extensive computational study was conducted using the following two solution approaches for two production scenarios:

1. Non-dominated Sorting GA II (NSGA-II) approach
    a. for a single-product ALBP
    b. for a mixed model ALBP
2. MOSA approach
    a. for a single-product ALBP
    b. for a mixed model ALBP

8.4.1 Application of NSGA-II for a single-product ALB

For the instance of a ten-task problem for a single-product assembly line, a solution approach based on NSGA-II was adopted. A population size of 200 and a mutation rate of 0.25 in 10 generations were considered. Using Pareto optimality, a set of equally effective, non-dominated solutions was obtained. Refer to the Tables, 8.8(a) to 8.8(c) for the non-dominated set of solutions generated by the model for the problem instance, and their corresponding objective function values are listed in Table 8.9. The chromosomes were non-dominated solutions obtained using NSGA-II, and a Pareto front is depicted in figure 8.3 for the single-product assembly line. This plot consists of Pareto fronts from each generation.

Table 8.8 (a). Chromosome 1

Workstation	WS_1		WS_2			WS_3			WS_4	
Task sequence	9	5	6	1	3	7	2	8	4	10
Worker sequence	4		1			6			5	

Table 8.8 (b). Chromosome 2

Workstation	WS_1		WS_2			WS_3			WS_4	
Task sequence	5	3	9	4	6	1	7	2	8	10
Worker sequence	1		2			3			6	

Table 8.8 (c). Chromosome 3

Workstation	WS_1		WS_2			WS_3			WS_4	
Task sequence	9	4	8	5	10	3	6	1	2	7
Worker sequence	4		3			2			6	

Table 8.9 Objective values of non-dominated solutions

	Objective 1	Objective 2	Objective 3
Non-dominated Solution 1	4	3	0.00191
Non-dominated Solution 2	4	4	0.00079
Non-dominated Solution 3	4	5	0.00026

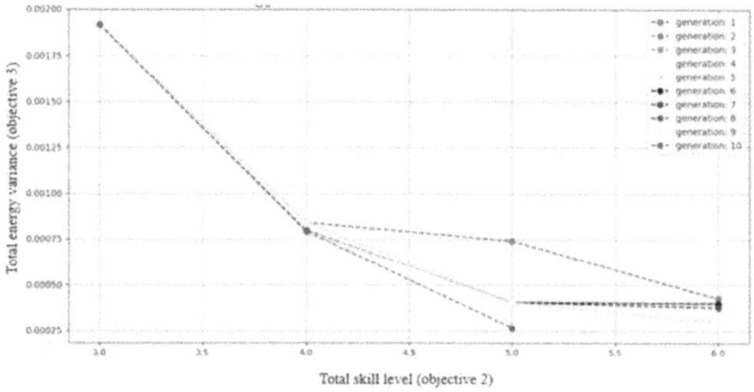

Figure 8.3 Plot of Pareto front using NSGA-II for a single-product problem instance

8.4.2 Application of NSGA-II for a mixed model ALB

All the parameters considered in the mixed model were the same as those in the single-product model, including the number of tasks, generations, and mutation rate. Using Pareto optimality, a set of equally good Pareto-optimal solutions is obtained. The non-dominated set of solutions generated by the model for the ten-task problem are provided in Tables 8.10(a) – 8.10(c), and their corresponding objective function values are listed in Table 8.11. The above-cited chromosomes are non-dominated solutions obtained using NSGA-II, and a Pareto front is depicted in Figure 4 for the mixed model assembly line.

Table 8.10 (a). Chromosome 1

Workstation	WS_1		WS_2			WS_3			WS_4	
Task sequence	8	4	3	9	5	1	6	7	10	2
Worker sequence	6		4			1			2	

Table 8.10 (b). Chromosome 2

Workstation	WS_1		WS_2			WS_3			WS_4	
Task sequence	4	6	1	2	9	3	7	8	5	10
Worker sequence	1		3			5			2	

Table 8.10 (c). Chromosome 3

Workstation	WS_1		WS_2			WS_3			WS_4	
Task sequence	9	4	1	2	8	6	5	10	3	7
Worker sequence	3		4			2			5	

Table 8.11 Objective values of non-dominated solutions

	Objective 1	Objective 2	Objective 3
Non-dominated Solution 1	4	6	0.0121667
Non-dominated Solution 2	4	7	0.0105847
Non-dominated Solution 3	4	8	0.0048779

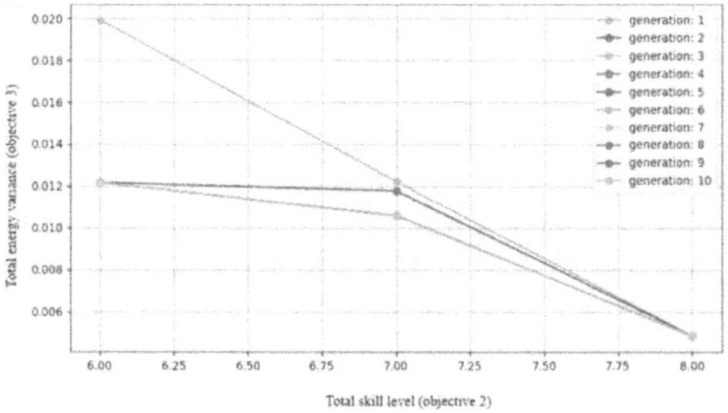

Figure 8.4. Plot of Pareto front using NSGA-II for a mixed model problem instance

After obtaining the results shown above, the task can be allocated to a workstation based on the task sequence, and the worker can be assigned to a workstation based on the worker sequence in an assembly line. From the Pareto chart for both single products, as shown in Figure 8.3, and multiple products, as shown in Figure 8.4, it can be inferred that both objective values decrease as the number of generations changes from 1 to 10. In the Pareto optimality approach, instead of obtaining only one solution, a set of Pareto-optimal solutions was obtained without assuming any weighting of the objectives. Hence, the Pareto optimality approach is considered superior to the weighted approach. This is a unique contribution of the present study to the multi-objective ALBP domain.

8.4.3 Application of MOSA for a single-product ALB

For the instance of a task problem for a single-product assembly line, a solution approach based on MOSA was adopted. The non-dominated set of solutions generated by the model for the hypothetical ten-task problem are provided in the Tables 8.12(a) – 8.12(c), and their corresponding objective function values are listed in Table 8.13 The chromosomes cited above are non-dominated solutions obtained using the MOSA, and the Pareto front is depicted in Figure 5 for the single-product assembly line. This plot consisted of a Pareto front from a single generation. The Pareto front consists of non-dominated solutions with total skill level (Objective 2) and total energy variance (Objective 3) values, as shown in Figure 8.5.

Table 8.12(a). Non-dominated Solution 1

Workstation	WS_1		WS_2			WS_3			WS_4	
Task sequence	5	3	9	10	4	1	8	7	2	6
Worker sequence	1		5			6			3	

Table 8.12(b). Non-dominated Solution 2

Workstation	WS_1		WS_2			WS_3			WS_4	
Task sequence	1	9	7	4	8	6	10	2	3	5
Worker sequence	4		6			5			3	

Table 8.12(c). Non-dominated Solution 3

Workstation	WS_1		WS_2			WS_3			WS_4	
Task sequence	9	1	4	7	5	3	10	8	6	2
Worker sequence	4		1			5			2	

Table 8.13. Objective values of non-dominated solutions

	Objective 1	Objective 2	Objective 3
Non-dominated Solution 1	4	2	0.000677
Non-dominated Solution 2	4	3	0.000150
Non-dominated Solution 3	4	5	0.000138

It is inferred that the total skill objective for the best solution is 2, which is the lowest among all the generated solutions. However, its energy variance objective is 0.000677, which is higher than that of many other solutions. The energy variance among the workstations that had the lowest value for the set of sequences represented in Table 8.13 was 0.000138; however, its total skill objective was higher at 5. The algorithm helps to identify the set of non-dominated solutions, and based on the requirements and priorities, a decision-maker can choose the best sequence for line balancing.

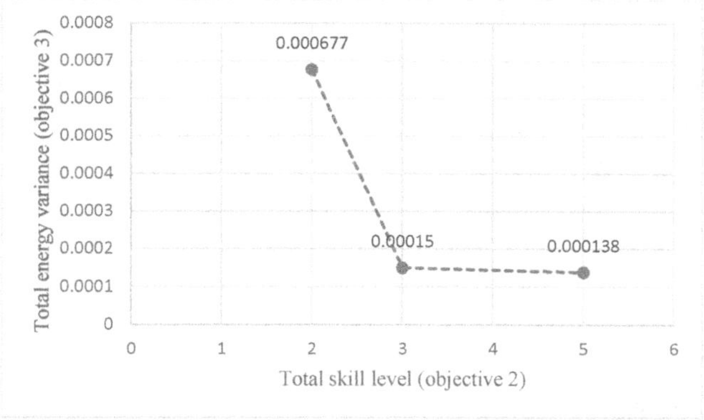

Figure 8.5. Plot of Pareto front using MOSA for a single-product problem instance

8.4.4 Application of MOSA for a mixed model ALB

For the problem instance of ten tasks for a mixed-model assembly line, a solution approach based on MOSA was adopted. The non-dominated set of solutions generated by the model for the hypothetical ten-task problem are provided in tables 8.14(a) – 8.14(d), and their corresponding objective function values are listed in Table 8.15. It can be observed from Table 8.15 that the lowest value of the total skill level, that is, for Objective 2, is 2 for a particular set of tasks, whereas the lowest value of the energy variance (i.e., for Objective 3) is in non-dominated Solution 4, as shown in Table 8.15. A Pareto front is plotted, which consists of non-dominated solutions with the total skill level (Objective 2) and total energy variance (Objective 3) values, as shown in Figure 8.6.

Table 8.14(a). Non-dominated Solution 1

Workstation	WS_1		WS_2			WS_3			WS_4	
Task sequence	9	1	5	7	6	10	8	3	2	4
Worker sequence	6		4			1			3	

Table 8.14(b). Non-dominated Solution 2

Workstation	WS$_1$		WS$_2$			WS$_3$			WS$_4$	
Task sequence	1	7	6	9	2	3	10	5	4	8
Worker sequence	3		6			2			4	

Table 8.14(c). Non-dominated Solution 3

Workstation	WS$_1$		WS$_2$			WS$_3$			WS$_4$	
Task sequence	8	7	9	4	1	5	6	3	2	10
Worker sequence	1		3			4			2	

Table 8.14(d). Non-dominated Solution 4

Workstation	WS$_1$		WS$_2$			WS$_3$			WS$_4$	
Task sequence	7	8	9	1	4	3	10	5	6	2
Worker sequence	5		3			2			4	

Table 8.15 Objective values of non-dominated solutions

	Objective 1	Objective 2	Objective 3
Non-dominated Solution 1	4	2	0.0647404
Non-dominated Solution 2	4	3	0.0136977
Non-dominated Solution 3	4	4	0.0116585
Non-dominated Solution 4	4	5	0.00460736

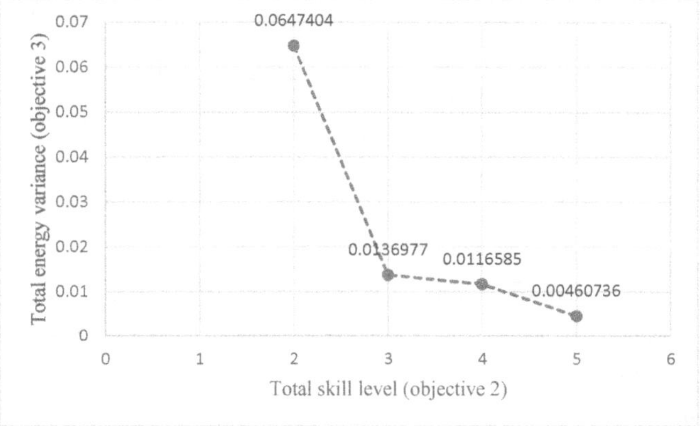

Figure 8.6. Plot of Pareto front using MOSA for a mixed model problem instance

8.5 Comparative analyses of the proposed solution approach NSGA-II and MOSA

For any solution approach, the computational time, or runtime, is crucial for indicating its efficiency. It has been reported that, for any combinatorial-type problem, the runtime increases exponentially as the size of the problem increases. In this computational study, the design of experiments was applied to identify the effect of input factors on the runtime of a solution approach in a mixed model assembly line.

In this study, five critical factors were identified that varied between the two levels, as shown in Table 8.16. Precedence constraints were expressed as a percentage of the number of tasks and were set at 10% and 20%. The skill levels were rated on 3- and 5-point scales. Similarly, the number of workstations was set to four and five, the number of tasks to 10 and 12, and the number of product models to six and eight.

Table 8.16 Factors and its levels identified for the design of experiments

Factors	Level 1	Level 2
Precedence constraints (% of task)	10	20
Scale adopted for rating skill levels	3	5
Number of workstations	4	5
Number of tasks	10	12
Number of models	6	8

For a comparative analysis of the proposed solution approaches in terms of computational time and runtime, an elaborate computational experiment was performed. The model was replicated three times for each run of the experiment, and the respective run times were recorded. Thus, $2^5 \times 3$ trials were performed. To investigate the influence of various factors, such as precedence constraints, skill levels, number of workstations, number of tasks, and number of products, on the two proposed solution approaches, each was applied to a set of 32 hypothetically generated problem instances at different combinations of the two levels of factors, as listed in Table 5. For each problem instance, both of the proposed approaches were applied and run three times. The computational time or runtime required for both of the proposed solution approaches in each iteration is listed in Table 8.17.

Considering the three iterations, it was found that the average runtime for NSGA-II was 37.484 s, and that for MOSA was 10.012 s. Therefore, it can be concluded that the MOSA computationally outperformed the non-dominated sorting GA (NSGA-II).

Table 8.17 Experimental design and the observed runtime for different combinations for three iterations

Precedence constraints (%)	Skill level (scale)	No. of workstations	No. of tasks	No. of products	MOSA			NSGA-II		
					Runtime (s)					
					Iteration 1	Iteration 2	Iteration 3	Iteration 1	Iteration 2	Iteration 3
-1	-1	-1	-1	-1	6.67	7.31	5.92	11.312	10.584	10.205
-1	-1	-1	-1	1	5.62	8.92	5.43	10.839	10.732	10.899
-1	-1	-1	1	-1	9.76	8.97	10.10	20.943	29.198	17.108
-1	-1	-1	1	1	10.58	9.89	10.27	22.104	21.654	21.461
-1	-1	1	-1	-1	12.59	11.19	10.28	13.260	12.140	12.418
-1	-1	1	-1	1	21.42	13.13	17.94	12.409	12.836	12.967
-1	-1	1	1	-1	12.43	9.38	8.27	21.34	23.346	25.242
-1	-1	1	1	1	5.48	5.58	6.80	27.464	29.296	27.915
-1	1	-1	-1	-1	6.98	5.93	6.12	12.166	11.930	11.967
-1	1	-1	-1	1	3.71	5.25	5.40	15.848	14.958	17.824
-1	1	-1	1	-1	8.04	8.72	4.08	22.392	25.454	25.029
-1	1	-1	1	1	13.98	14.94	18.54	31.119	38.497	41.626
-1	1	1	-1	-1	13.33	9.73	11.10	20.206	15.693	17.336
-1	1	1	-1	1	14.61	9.19	11.65	16.829	19.606	17.666
-1	1	1	1	-1	7.63	7.95	10.71	48.286	44.414	44.263
-1	1	1	1	1	7.99	8.36	9.43	44.595	53.620	47.770
1	1	-1	-1	-1	5.45	6.85	5.97	15.037	21.280	20.128
1	1	-1	-1	1	5.79	3.66	5.29	14.217	14.509	13.848
1	-1	-1	1	-1	11.00	10.51	7.81	45.700	51.753	56.784
1	-1	-1	1	1	9.99	12.6	9.79	44.248	43.517	49.422
1	-1	1	-1	-1	9.28	8.97	9.15	16.863	19.662	17.618
1	-1	1	-1	1	19.14	17.81	17.98	16.636	18.328	17.519

Table 8.17 (Contd.)

Precedence constraints (%)	Skill level (scale)	No. of workstations	No. of tasks	No. of products	MOSA			NSGA-II		
					Runtime (s)					
					Iteration 1	Iteration 2	Iteration 3	Iteration 1	Iteration 2	Iteration 3
1	-1	1	1	-1	11.47	9.20	9.32	68.988	89.020	77.791
1	-1	1	1	1	10.12	12.40	14.14	79.663	75.415	78.221
1	1	-1	-1	-1	5.51	6.41	6.18	16.251	16.520	14.600
1	1	-1	-1	1	4.88	6.88	5.75	21.682	19.457	19.586
1	1	-1	1	-1	5.68	9.72	5.90	56.867	57.499	60.510
1	1	-1	1	1	16.40	14.11	15.34	69.194	78.047	89.787
1	1	1	-1	-1	12.72	15.11	15.37	24.668	24.341	25.762
1	1	1	-1	1	11.36	10.82	13.83	26.905	24.155	27.257
1	1	1	1	-1	13.98	9.95	10.64	135.62	134.18	143.96
1	1	1	1	1	14.51	14.18	15.02	138.81	162.26	157.63
Average runtime					10.013			37.484		

8.5.1 Performance analyses

Furthermore, the Pareto charts of the standardized effects of the factors listed in Table 5 on the proposed solution approaches, NSGA-II and MOSA, are shown in Figures 8.7 and 8.8 respectively. With reference to NSGA-II, it was observed that the number of tasks had the greatest influence on the runtime of the solution approach, followed by the percentage of precedence constraints and the interaction effect of both. Therefore, it is not advisable to apply NSGA-II as a solution as the number of tasks increases. Similarly, in situations with a higher precedence relative to the number of tasks, NSGA-II requires more computational time.

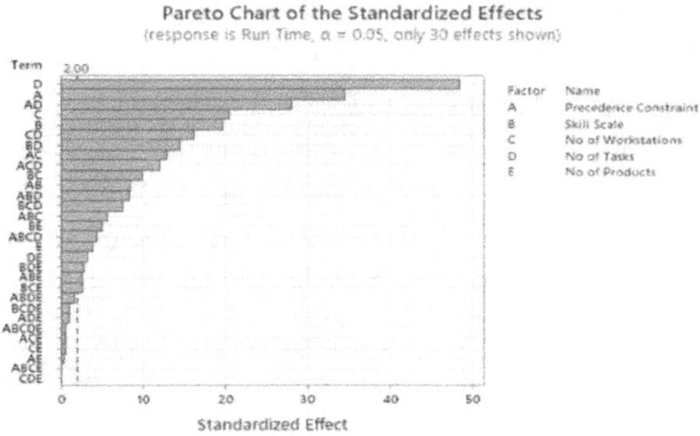

Figure 8.7. Pareto chart of the standardized effects of factors on NSGA- II

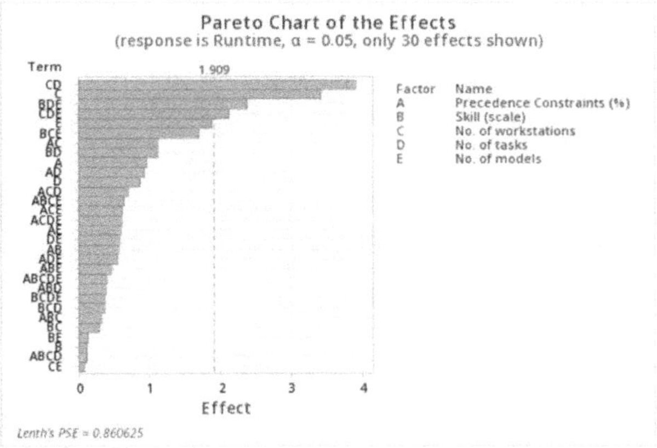

Figure 8.8. Pareto chart of the standardized effects of factors on MOSA

Moreover, with reference to MOSA, it is observed that the interaction effect of the number of workstations and number of tasks has the greatest influence on runtime, followed by the number of workstations, and then by the interaction effect of skill level, number of tasks, and number of products/models. However, the MOSA performed relatively better in both production scenarios. Furthermore, to investigate the effect of individual factors on the runtime of the solution approaches, NSGA-II and MOSA, a main effects plot was analyzed, as depicted in Figures 8.9 and 8.10.

From Figure 8.9, it can be inferred that the number of tasks has the greatest influence on the runtime of NSGA-II. Similarly, with reference to Figure 8.10, the number of workstations has the greatest influence on the runtime of the MOSA. The runtime of the solution approach increases when the various factors considered increase, except for the number of products in NSGA-II and the skill levels in MOSA.

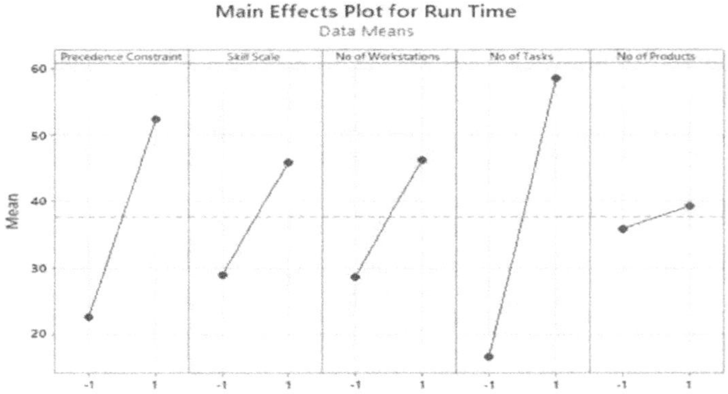

Figure 8.9. Effect of individual factors on runtime of NSGA-II

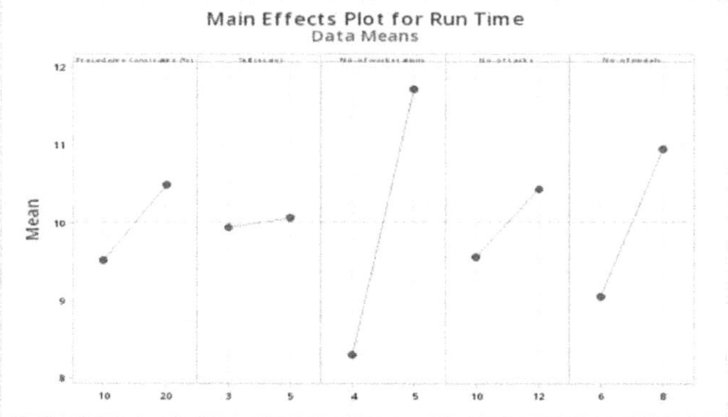

Figure 8.10. Effect of individual factors on runtime of MOSA

The interaction plot of the factors helps to interpret the main effects between the lines. Therefore, the interaction plots of the factors considered with respect to the solution approaches of NSGA-II and MOSA are shown in Figures 8.11 and 8.12 respectively. When analyzing these interaction plots, if the lines are not parallel to the x-axis, it can be inferred that there is a main effect between the factors considered; that is, across all factor levels, the response mean is not the same. The steeper the slope of the line, the greater the magnitude of the main effect. According to Figures 8.11 and 8.12, a greater magnitude of the main effects is more evident in the case of the MOSA than in the NSGA-II. This proves that the solution approach of the MOSA performs better than NSGA-II in solving a multi-objective ALBP.

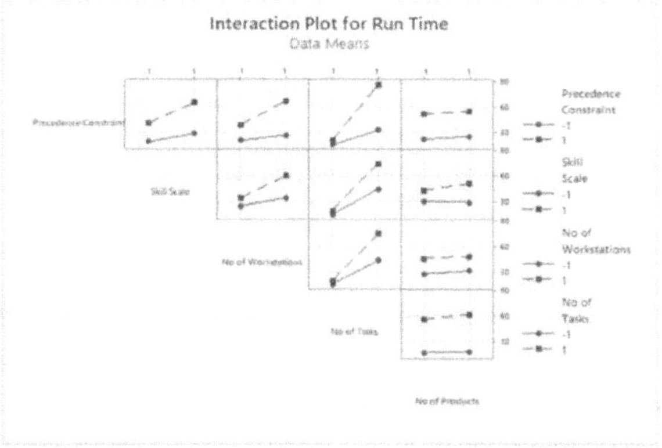

Figure 8.11. Interaction plot for runtime of NSGA-II

The results illustrate how the algorithm simultaneously minimises the number of skilled personnel and energy load variance, thereby achieving the proposed objectives. The algorithm was tested using real-time data obtained from the assembly experiments. The algorithm helps minimise the number of workstations and simultaneously minimises the number of skilled workers and the energy variance among workstations. Workers are less strained because their physical capabilities are considered when assigning tasks to the workstations. This maximizes the productivity of the firm and distributes the workload equally among the workstations.

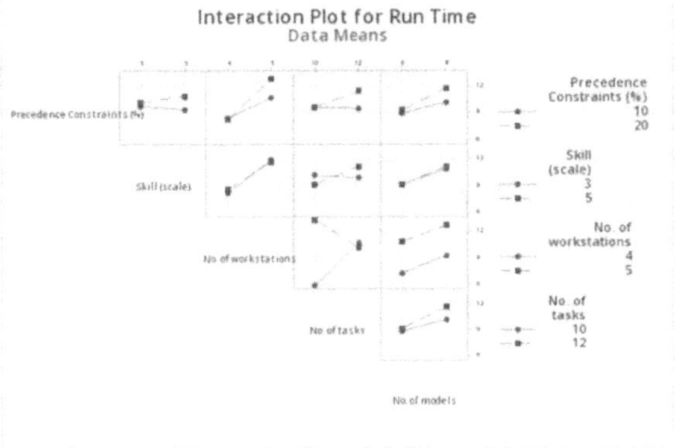

Figure 8.12. Interaction plot for runtime of MOSA

8.6 Discussion

A multi-objective ALBP is addressed in this work by proposing NSGA-II and MOSA approaches to minimise the overall skill level required and variance in workers' energy expenditure across workstations. Mehta [120] reported the need to integrate both the physical and cognitive aspects of ergonomics, as most of the work environment is involved. However, the integration of these aspects is limited in the literature.

The approach described in this research combines physical (e.g., energy expenditure) and cognitive ergonomics (e.g., worker skill). Solving these objectives of the mathematical model contributes to an improved workstation design by evaluating tasks to make them compatible with people's needs, abilities, and limitations. Furthermore, a comparative analysis of NSGA-II and MOSA was conducted in single- and mixed model production scenarios to identify the dominant algorithm. The computational analysis indicated that the runtime performance of MOSA was 73.287% better than that of NSGA-II; therefore, MOSA outperformed NSGA-II. In summary, this research helps improve the productivity and efficiency of workers with proper planning and allocation by solving the ALBP.

8.6.1 Key findings

The proposed approach will aid in identifying worker skills to determine the right individuals to allocate to a particular workstation, considering the complexity of the task and also reduces the cognitive workload. In the existing literature, researchers typically employ a weighted-sum approach [9] to address multi-objective problems in line balancing. However, the drawback of the weighted-sum method is its reliance on the assigned weights, where even a small variation can significantly alter the objective function values. This study addresses this drawback by utilizing the Pareto optimality method. It contributes to determining the optimal workplace arrangement by examining factors such as workstations, skills, and energy expenditure through an iterative process aimed at identifying the most efficient and ergonomic layout.

Overall, this study is an effective approach to the design of ergonomic processes, as it allows the production units to optimize multiple objectives simultaneously, leading to a more efficient and ergonomic assembly line design.

8.6.2 Practical implications

The practical implications of diesel pump manufacturing unit, random worker allocation in the production units of small scale and large-scale industries and how ALBP improves ergonomics are discussed in this section.

8.6.2.1 Diesel pump manufacturing unit

This study considers the assembly line of a diesel pump manufacturing unit in India. As explained above, a wide variety of small parts and end products exist. Manufacturing units experienced a decline in productivity, and it is essential to identify the reason for this decline. However, breaking down the finished product into subparts and identifying the root cause are time-consuming for the operations team. After using scientific techniques, such as the Shainin method, the operation team identified issues with the barrel and plunger. This study assessed the assembly lines for barrels and plungers, considering six types of barrel-plunger assemblies. The balls inside the assembly were interchanged, and this was identified as an issue. In one product, it was assumed to be a steel ball, and in the other, it was a ceramic ball. This problem has been studied, and it has been found that workers require proper cognitive abilities. In addition, some workers could not distinguish between the colors or weights of the balls. Thus, in the present study, some inputs were obtained from a previous study by Giridhar and Panicker [112] on workers' cognitive skills to solve the ALBP. Kindly refer to the figure 8.12 for a comparison between the existing work by Giridhar and Panicker [112] (i.e., Chapter 4,5 and the present work). Using this intervention, the results from ALBP increased the first-pass yield from 55% to 89%, which helped reduce rework. This underscores the practical implications of the study. The proposed allocation method can be employed in special processing industries that use stainless steel and carbon steel to avoid weld defects and reduce material waste.

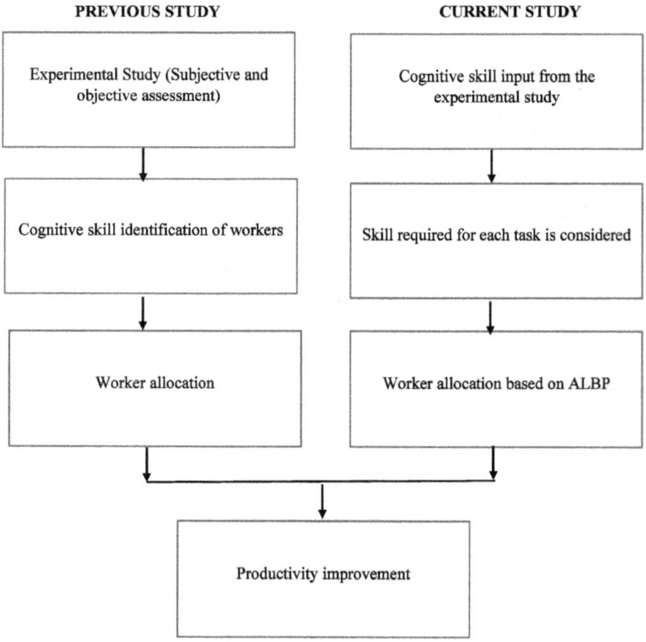

Figure 8.12 Comparison between Chapter 4, 5 and this study

The methodology employed in this study was applied to all the production units with certain modifications. The challenge in this process is to identify a problem in an actual scenario and use it to solve it. In most cases, this problem can be resolved using experienced personnel. The problem must be converted from a subjective form into a mathematical formulation, referred to as the ALBP. One of the significant limitations of the present study is that, in each industry, the ALBP formulation is different, and the selection of the right method is crucial after defining the problem.

8.6.2.2 Random worker allocation in the production units of small-scale and large-scale industries

In manufacturing industries, the random assignment of workers to certain tasks leads to inefficiency, resulting in a waste of time, resources, and the delivery of finished products to customers. The user's perceived level of mental effort, known as the cognitive workload, is influenced by various factors, particularly task load and task design. In general, the greater the skill, the lower the cognitive workload. However, there are certain exemptions when considering age, gender, etc., which are separate topics for discussion. The main reasons for the random allocation of workers by the employer are lack of skilled workers, lack of knowledge of ALB, cost constraints, space constraints, workers leaving the company for better incentives and recruiting temporary workers (on a contract basis) to reduce costs who are not skilled.

Although random allocation cannot be solved easily, the present study suggests that employers must determine the skills of workers and assign them to appropriate workstations. The proposed approach, in turn, lowers training costs and enables the training of suitable workers. Appropriate training equips individuals with skills that can be leveraged to improve their employers' overall output. This study contributes to small- and medium-sized (SME) enterprises that can apply the proposed method to their units. Moreover, according to the recommendations of ISO 26000 standards, improving wellbeing at work is a core component of social responsibility. Therefore, firms must optimize their workload in assembly lines by considering ergonomics.

8.6.2.3 How ALB improves ergonomics

ALB is used to reduce waste and increase the efficiency of an assembly line. When an assembly line is balanced, each workstation is designed to have an equal workload, and each worker performs the same amount of work. Hence, this prevents the overburdening of any worker and reduces the physical stress. ALB can improve worker posture by designing workstations that accommodate different worker heights and postures. Workload balancing also helps to minimise repetitive motion and reduces the likelihood of injuries, such as tendinitis and carpal tunnel syndrome. This improves the worker satisfaction, thereby resulting in a more enjoyable work experience. Balancing the assembly line renders the process more efficient, reducing wasted motion and downtime which reduces productivity.

Efficient assembly lines require less lifting, stooping, bending, and twisting, which leads to fewer musculoskeletal disorders and significant ergonomic improvements. In conclusion, ALB can significantly improve ergonomics and result in a safer and healthier environment for workers. The methodology used in this study can serve as a guide for small- and medium-scale industries. This study can also be applied to large production units, provided that they start by splitting the large production units into small workstations and then expand the study to the whole unit.

8.6.3 Study limitations

Only an expert in this area, who has knowledge of ALBP and ergonomics, can extract the exact data from an industrial environment and solve the problem. This is a time-consuming process that requires significant attention to obtain valid and reliable data. With regard to ergonomics, anthropometric data vary for different nationalities, and their energy expenditures vary based on the tasks that lead to a change in the results. Thus, with the methodology in this study, the modeling part can be performed, but a variation in the results is expected. The influence of ongoing technological developments such as digitization, autonomous robots, and Internet-of-Things (IoT) technology and their adaptation to ALBP are not considered in this study. The varying ALBP models that are required for the different phases of the lifecycle of an assembly line and the products it manufactures were not considered in this study. The present study mainly focused on the initial line balance. Combined models that integrate the ALBP with related decision tasks are also a limitation of this study.

8.6.4 Areas for further research

In the future, we will consider job rotation when allocating tasks to the workers. Similarly, the application of collaborative robots should be considered for use in assembly lines to reduce energy expenditure and enhance worker ergonomics. The collaborative (human-robot collaboration) ALBP (C-ALBP) can be performed with automated assembly lines to identify the feasibility of expansion to improve flexibility and efficiency in assembly systems.

As a main future research topic, it is proposed to carry out an accurate analysis in terms of the obtained ALB between workstations by trying different types of ALBP. In addition, the number of tasks, product variety, experimental parameters, and anthropometric data can be varied. Furthermore, when we consider a task requiring a greater physical workload, the energy expenditure will vary, thereby influencing the results. This changed the number of iterations required to obtain a near-optimal solution.

Some areas of future research include adapting the problem for mixed-model assembly lines and different line configurations, considering the learning effect of workers, addressing a production environment in which task times are stochastic, and optimizing different performance measurements. In addition, the problem can be solved using different heuristic and meta-heuristic methods, and comparisons can be made between them.

8.6.5 Ergonomics Improvement in assembly lines

Ergonomic improvement in assembly lines is crucial for the safety and wellbeing of workers as well as the overall efficiency of the assembly process. Some strategies for improving ergonomics in assembly lines are task analysis, workstation design, material handling, automation and training.

Analyze the tasks that workers perform and identify potential ergonomic hazards. This can help determine adjustments to work methods, equipment, and workstation design. Workstation design plays a significant role in assembly line ergonomics.

Workstations should be designed to be adjustable and to accommodate different user heights and postures. This can involve adjusting the height of the equipment, using padded flooring, or providing appropriate lighting. Material handling is an integral part of the assembly line. The design of material handling equipment should be based on ergonomic principles with appropriate handling guidelines, techniques, and tools.

Automation can reduce manual labor and improve ergonomics in assembly lines. Automating repetitive and physically demanding tasks can reduce the risk of musculoskeletal disorders, lower the physical stress on workers, and allow them to focus on more complex tasks. Providing ergonomic training to workers can increase their awareness of potential hazards and assist in avoiding them. Additionally, workers can learn how to adjust their work posture and take breaks as needed, which can significantly reduce the risk of ergonomic injury. Overall, ergonomic improvements in assembly lines can lead to significant benefits, including a reduced risk of injuries, improved productivity, and enhanced worker satisfaction.

8.7 Summary of Chapter 8

The application of ergonomic aspects in assembly lines has received little attention in the literature, although considering these aspects improves productivity and improves ergonomics in assembly lines. This study focused on a multi-objective ALBP with three objectives: minimisation of the number of workstations, total skill-level requirements of workers, and variance in the energy expenditure of workers among workstations. A literature review verified that many researchers have extensively investigated multi-objective ALBPs using the weighted-sum approach without considering the ergonomic aspects. To address this gap in the literature, this study proposes an approach that considers two algorithms, namely, NSGA-II and MOSA, by applying Pareto optimality.

A comparative analysis of the NSGA-II and MOSA in single- and mixed model production scenarios was conducted. Furthermore, a computational study with varying factors was performed to identify the dominant algorithm that can solve the multi-objective ALBPs, and the results demonstrated that the computational efficiency of MOSA was 73.287% better than that of NSGA-II.

This study minimises the number of workstations, total skill level required for workers on the line, and energy load variance among workstations. The energy load was evaluated by comparing the energy expenditure while performing tasks with the worker's physical energy limit, which was calculated based on anthropometric and physiological factors.

The results show that could assign workers to workstations while reducing the energy load variance under all conditions while considering skill constraints and energy expenditure. Even if workers have a high workload or are unable to perform assigned tasks, the algorithms for solving line-balancing problems that ignore ergonomics may generate solutions that minimise the number of workstations in an assembly line. Consequently, the recommended algorithms are well-suited for industries in which assembly activities necessitate dexterity and manual skills gained through experience. In such production units, skilled workers are valuable assets, like any other resource. The recommended algorithms can be used to design assembly plans based on specific individuals available to the organization, thereby preventing workers from being overloaded and stressed. The next chapter summarizes the major findings, contributions to the literature, limitations in the present research, and the scope for further research.

Chapter 9

Conclusion

9.1 Introduction

This chapter summarises the findings from the experimental studies pen assembly task, virtual cognitive task, studies with wireless EEG headset, and ALBP. The studies and methodologies used are both relevant to the production industry. The summary of the five studies of the present research is given here, which are detailed in Chapters 4 to 8. In addition to the summary, the chapter revisits the significant contributions of the research. The limitations of this study and the potential for future research are discussed in this section.

9.2 Summary of the research

A variety of literature sources were examined in the area of manufacturing focussing on topics associated with manual assembly, CLAM and ALBP. Additional topics cognitive abilities, methods to measure cognitive load are also identified. As manufacturing and manual assembly encompass a wide range of research topics, there are some topics that remain unexamined. With a preliminary review of commonly used factors in manufacturing and manual assembly, it was possible to explore different concepts within the literature review, which helped to get a thorough comprehension of the fundamental research domains. The experimental studies (pen assembly task and virtual cognitive task) mimicked the industrial environment and a methodology is created to allocate the workers based on their cognitive abilities and performance. The results of the studies done with subjects can be applied to the workers in a manufacturing unit.

9.2.1 Influence of cognitive aspects of information and material presentation in a manual assembly environment

This research presents the theme of cognitive aspects to be considered in a manual assembly and identifying the worker performance, skills, and thereby allocating workers to the workstations. The study depicts how significantly worker performance in manual assembly systems differs from what is typically accounted for by workstation designers and how such differences might be linked to cognitive aspects. The significant differences found in worker performance are due to the varying cognitive capabilities of the workers. The influence of cognitive aspects, including information presentation, and material presentation on worker performance is identified. Poor workstation design of manual assembly processes will increase cognitive load, resulting in reduced productivity. To address this issue, a real-time case study is proposed in a pen-recycling unit with a methodology. The methodology aims to identify the right worker for the right workstation through a pilot study. This approach is simple and economical and can be applied to all manual assembly workstations with a wide variety of products.

A factorial experiment design with eight scenarios is created with combinations of information and material presentation. The pen assembly task is done by thirty subjects, and the number of pens assembled correctly in each scenario is recorded. This pen assembly experiment serves as an analogy to the pen recycling unit in Kerala. The scenarios in the experimental study are designed to resemble the workstations in the pen recycling unit and the performance of the subjects can be used to allocate the workers. Workers are allocated to appropriate workstations based on their performance analysis and required cognitive abilities identified. The maximum number of pens assembled in Scenario 8 (with part-numbered boxes and a photograph) is 148, while the minimum number of pens assembled is in Scenario 2 (grouped parts with instructions), which is 4.

Thus, the necessity for a proper information presentation and material presentation in a workstation to improve worker efficiency is recognised.

9.2.2 Identifying the cognitive abilities using virtual cognitive test

In this study, eight cognitive abilities such as perception, attention, memory, motor skills, language, visual and spatial processing, executive functions and learning are considered in the virtual cognitive task. For testing each cognitive ability, brain teasers are used, and the thirty subjects are rated accordingly. The weights of these cognitive abilities are varied, and thus the influence of these abilities on the total score of the subject is analysed using statistical analysis. It is inferred that the rankings remain unchanged even when different weights are assigned to the cognitive ability scores. Subjects who have the best cognitive ability and subject who has the least is identified. The fixed score of 73 from brain blog research websites needs to be statistically tested and need to identify whether the rankings will be the same when the cognitive ability scores are given different weights. This suggests that the cognitive ability test conducted is valid and can be used for further studies and the weightages given are appropriate. A subject must try to develop his/her skills which he/she lacks to improve the overall performance. Feedback of the score in each section can be made available and can train the subjects for respective tasks. Subject 30 (56 out of 73) has the highest score and subject 5 (32 out of 73) has the lowest. Cognitive abilities identification helps to fit the workers in the right workstation.

9.2.3 Worker allocation using personalised models from cognitive abilities

This study combined the concepts of cognitive load, cognitive abilities, and worker allocation, and a methodology is framed. The technique is a personalised model, and it varies on the industrial environment and other factors such as the size of the industry, workers, workstations, etc. The performances of subjects in pen assembly task and virtual cognitive task are combined to build personalised models based on the cognitive abilities required to perform the assembly task. Statistical methods, namely, Poisson regression and omnibus test, are applied. It is observed that visual perception, memory, visual and spatial processing, executive functions and learning are the cognitive abilities required to perform the pen assembly task. The weightage of each cognitive ability for best performance is identified and based on that worker allocation is done. Out of the thirty subjects only six of them were eligible to work in the assembly task considered to provide better productivity. In practical situations, this will not be possible since the employers will have to incur losses, a novel methodology is proposed wherein all the workers can be allocated based on their cognitive abilities. With effective allocation, only one worker requires training which further reduces the cost.

The eight scenarios are grouped according to complexity into four tasks. The total number of products assembled correctly in task 1 and task 2 determines whether the subject can be good at Quality control. This means that when it comes to high and medium complexity conditions, worker can disassemble and rework without any errors The total number of products assembled correctly in task 3 and task 4 determines whether the subject can be good as an Assembler. Task 3 and task 4 signifies that when it comes to assembly, only with proper information and material presentation the errors can be reduced. Out of the 30 workers, only 20% of them can be assembler and 6.67% can be in quality control. When the skills of workers are identified based on studies, training can be provided depending upon the complexity of the task to improve worker efficiency.

9.2.4 Investigation of pen assembly worker performance using wireless EEG headset

A wireless EEG headset is used to validate the pen assembly task. The pen assembly task is repeated by using the wireless EEG headset. An increase in cognitive load is measured by an increase in brain activity, and the variation is identified for each scenario. The experiment with the wireless EEG headset revealed that the observed differences in band power among the different scenarios are sensitive to changes in cognitive load. When the amplitude of these frequency bands increases, it suggests a higher cognitive workload, indicating that tasks in scenarios 1, 2, 5, and 8 were more demanding and complex for the thirty subjects.

According to the pen assembly experiment, Scenarios 6 and 8 have the least number of errors and Scenarios 4,7 have the greatest number of errors. The number of pens assembled correctly is highest for Scenarios 8 and 6 and Scenarios 5 and 2 have the least. The experiment with a wireless EEG headset revealed that the observed differences in Theta and Alpha band power among the different scenarios indicate that these average band powers are sensitive to variations in cognitive load. A higher cognitive load was seen in scenarios 1, 2, 5, and 8 which were more demanding and complex for the participants. Scenario 6 (part numbered box with instructions) has the lowest cognitive load. This suggests that the combination of clear visual cues (part numbered box) and accompanying instructions reduced the mental effort required by participants. The total number of pens assembled without errors is 46.28 %, whereas the number of errors made is 53.71 % considering 808 pens assembled by 30 subjects. Photograph as an information presentation gives better results, but for scenario 8, the photograph created confusion and the cognitive load was increased. This again supports the importance of proper material and information presentation to improve worker productivity.

9.2.5 Effect of ergonomic aspects on single and mixed model assembly-line balancing problems

Despite the potential benefits, the literature has largely overlooked the integration of ergonomic considerations in assembly lines. Considering these aspects improves productivity and improves ergonomics in assembly lines. This study focused on a multi-objective ALBP with three objectives: minimisation of the number of workstations, total skill-level requirements of workers, and variance in the energy expenditure of workers among workstations. A review of the literature verified that many researchers have extensively investigated multi-objective ALBPs using the weighted-sum approach without considering the ergonomic aspects. To address this gap in the literature, this study proposes an approach that considers two algorithms, namely, NSGA-II and MOSA, by applying Pareto optimality.

A multi-objective Assembly Line Balancing Problem (ALBP) is investigated, with the aim of minimising the overall skill level required and the variance in the energy expenditure across workstations. This study proposes two solution approaches, namely Non-dominated Sorting Genetic algorithm II (NSGA-II) and Multi-Objective Simulated Annealing (MOSA), using Pareto optimality. A comparative analysis of the NSGA-II and MOSA is conducted in both single and multi-product production scenarios. The computational analysis indicates that the runtime performance of MOSA is 73.287% better than that of NSGA-II; therefore, MOSA outperforms NSGA-II.

9.3 Contributions of the research

In summary, cognitive aspects need to be considered while designing workstations, as they influence worker performance. Information and material presentation in manual assembly should focus to reduce the cognitive load on workers. An appropriate instruction manual and an ergonomically designed workstation, considering both physical and cognitive aspects are essential. The proposed algorithms solved ALBP considering ergonomic aspects. This research provides a methodology for allocating the workers based on their cognitive abilities. The proposed worker allocation methodology in this study can be used by micro, small and medium enterprises.

The following are the main contributions and highlights of the present investigation:

- Cognitive aspects of information and material presentation in a manual assembly improves worker efficiency and reduce errors – Guide to the employer for considering both physical ergonomics and cognitive aspects while designing the workstation
- Personalized models based on cognitive abilities for worker allocation.
- A methodology to implement the cognitive aspects and identifying the cognitive abilities of workers in a manual assembly unit.
- Development of meta heuristic algorithms using ergonomic aspects in ALBP

9.4 Limitations of the research

The limitations of this study are listed as follows:

- Traditionally, cognition has been viewed as mental activities occurring within the human brain, with cognitive abilities that allow individuals to perceive and engage with the world. Every day, humans engage in cognitive activities such as perception, decision-making, problem-solving, and memory processes. Despite the comprehensive nature of human cognition, it has limitations. One such limitation arises when the cognitive system is exposed to stimuli, resulting in what is commonly known as a cognitive load. The cognitive load varies between workers at all times and is influenced by the context, the tasks at hand, and the demands that these tasks place on each person. External factors that contribute to this extra cognitive load are not taken into account.
- The amount of information to be processed, delays, quick decisions, a larger variety of products, and the physical layout of the workstations can be handled easily if time pressure is not considered, but when combining these with time pressure, the cognitive load will vary.
- Heart rate variability can be used to predict stress at work and other health factors [121]. However, there are mixed opinions about heart rate variability and whether it can be considered as a valid method [122] to measure cognitive load.
- Only an expert in this area, who has knowledge of ALBP and ergonomics, can extract the exact inputs from an industrial environment and solve the problem. This is a time-consuming process that requires significant attention to obtain a valid and reliable output.
- With regard to ergonomics, anthropometric data vary for different nationalities and their energy expenditures vary according to the tasks that lead to a change in the results. Thus, with the methodology in this study, the modelling part can be performed, but a variation in the results is expected.
- In this study, the influence of ongoing technological developments such as digitization, autonomous robots, and Internet-of-Things (IoT) technology and their adaptation to ALBP are not considered.

9.5 Scope for further research

Each of these topics presents unique research gaps that require a multidisciplinary approach, including cognitive aspects, CLAM factors, cognitive abilities, ALBP, and worker allocation.

The present study can be extended further in the following ways:

- In future studies, the method could be improved by applying this methodology to workstations that use tools, robots, etc. Rather than manual assembly tasks, machine and workforce reconfigurations can be considered in a reconfigurable manufacturing system.
- Field studies incur high costs and require considerable effort, along with obtaining industrial permissions. In contrast to many laboratory studies, it is recommended that further research examine the findings of this thesis in real-world contexts.
- Skin conductance, that is a proven method which helps evaluate mental load can be used in future studies. Mehler et al. [123] investigated mental workload in a simulated driving environment with heart rate and skin conductance (sweat gland activity) as its main focus measures. Even though the equipment is costlier and causes uneasiness to the user, this can be used as a method to evaluate mental load.
- Eye tracking have been applied to measure mental workload in recent years. This technology typically relies on real-time video recordings of the eye captured by cameras mounted on the head of the subject or located on a screen in front of the subject [124]. Future studies can utilise Eye tracking methods.
- The application of collaborative robots for use on assembly lines should be considered to reduce energy expenditure and improve worker ergonomics. Collaborative ALBP (human-robot collaboration) (C-ALBP) can be performed with automated assembly lines to identify the feasibility of expansion to improve flexibility and efficiency in assembly systems.

www.ingramcontent.com/pod-product-compliance
Lightning Source LLC
LaVergne TN
LVHW010721061126
851857LV00014B/1291